Innovations in Everyday Engineering Materials

T. DebRoy • H. K. D. H. Bhadeshia

Innovations in Everyday Engineering Materials

 Springer

T. DebRoy
Department of Materials Science
and Engineering
Pennsylvania State University
University Park, PA, USA

H. K. D. H. Bhadeshia
Department of Materials Science
and Metallurgy
University of Cambridge
Cambridge, UK

ISBN 978-3-030-57614-1   ISBN 978-3-030-57612-7   (eBook)
https://doi.org/10.1007/978-3-030-57612-7

This Springer imprint is published by the registered company Springer Nature Switzerland AG
The registered company address is: Gewerbestrasse 11, 6330 Cham, Switzerland

Preface

Many successful materials and processes are so good that people do not need to know of the tremendous ingenuity in their creation and application. Innovations made in artificial materials have been important in improving the quality of life. Many of the innovations originated from organised research although serendipity in creative settings has contributed to advancements.

Our intention with this book is to introduce a few of the key innovations without which our quality of life would be dramatically different. But in addition, we selected a couple of cases where we felt that it is realistic to expect quite significant developments. One such case is low-density steels and the other, high-entropy alloys.

More than anything else, we tried hard to keep this book small. Our goal was to create a book that was not too onerous to read, accessible to a wide audience and illustrate the vitality of the field of engineered materials. The book can, we think, be read in a leisurely fashion within a couple of days and yet can provide a degree of learning.

We hope that the book brings some balance to the trend where research is publicised before it has delivered. The greatest achievements illustrated in the examples come from scientists and engineers who take pleasure in creating materials that make a difference, no matter how long it takes or how many barriers have to be surmounted.

We gratefully acknowledge the support of Professors Kwadwo Osseo-Asare, Thomas W. Eagar, Ronald M. Latanision, Chobi DebRoy and David G. C. Robertson for their knowledge, patience and many valuable comments. Drs John Elmer, Jared Blecher, James Zuback and Tuhin Mukherjee made valued comments.

Pennsylvania, PA, USA   T. DebRoy

Cambridge, UK   H. K. D. H. Bhadeshia

Contents

Chapter 1
A Remarkable Innovation in Stainless Steel Making

1.1 Engineered Materials

An engineered material is one that is not found occurring naturally anywhere on earth. The adjective also implies that the material has some use. Imagine a material that is strong and tough, has great looks, costs about the same as a cup of coffee on a weight comparison, which relies on an atomic surface that fends away the environment. This is steel that will not stain.

The lure of their attractive appearance, their excellent properties and right price have all contributed to their many uses in large iconic structures (Fig. 1.1), buildings, bridges, automobiles, sophisticated kitchen appliances, ordinary kitchen utensils and medical devices. The latest *Tesla* truck has a stainless steel body, rather like the *DeLorean* that featured in the movie "Back to the future". All of these rely on a thin layer of chromium oxide that forms spontaneously on the surface. It is coherent, adherent and regenerates when scratched. The film protects the underlying steel from corrosion, hence the designation "stainless". It was in 1913 that Harry Brearley invented the steel in Sheffield, UK. However, the story here is about a young engineer Krivsky [1], fresh from college, pursuing fundamental research in a large corporate laboratory in a town that most people visit for a different purpose—Niagara Falls. What started as an unusual research result did not end up in the usual equations, charts and research papers. After a long saga filled with many formidable difficulties and uncertainties, the payback was a new process so technologically advanced and profitable that it disrupted the contemporary method of manufacturing stainless steel which quickly became obsolete throughout the world [1].

Like most major inventions looked at with hindsight, the process modification is in essence simple. Oxygen must be blown through the molten metal to reduce and control the concentration of carbon, but chromium has a strong affinity for oxygen and is an expensive element to lose as an oxide. Instead, Krivsky used a mixture of argon and oxygen, distributed uniformly inside the melt, thereby dramatically limiting the loss of chromium.

© Springer Nature Switzerland AG 2021
T. DebRoy, H. K. D. H. Bhadeshia, *Innovations in Everyday Engineering Materials*,
https://doi.org/10.1007/978-3-030-57612-7_1

Fig. 1.1 (**a**) Cloud Gate, nicknamed the *bean*, designed by Sir Anish Kapoor, built using 168 stainless steel plates joined together. The concept was inspired by liquid mercury. (**b**) An aesthetically pleasing stainless steel street lamp in Busan, South Korea. (**c**) Double-walled cups and saucers made of stainless steel. (**d**) An island restaurant made using stainless steel, in the River Mur in Graz, Austria

1.2   Unexpected Results

A metallurgist by training, Krivsky, had just joined the Metals Research Laboratories of the Union Carbide Corporation in Niagara Falls, New York after his doctoral work. His first job was to examine the high temperature metallurgical reactions between oxygen and hot molten alloys containing iron, chromium and carbon from which stainless steels are made. Professor Richardson in England and Drs Hilty and Crafts in the USA were all established researchers in this field, but their thermodynamic data on the iron–chromium–carbon system were somewhat different. Dr Krivsky was trying to resolve these discrepancies. We shall see that working with hot liquid steels has its challenges in a laboratory, but even more so when scaling to industrial production.

Results of Krivsky's initial studies on blowing oxygen through the melt were complicated by the heat of reaction when carbon and chromium combined with oxygen and this must have dogged the earlier experiments; to have confidence in the scientific data, it is necessary to control the temperature. Krivsky decided to use oxygen diluted with an inert gas, argon, in order to slow down the rate of oxidation and hence to better control the temperature. This appeared to be a rock solid strategy, because argon, as an inert gas should not interfere with any reaction and only slow down the exothermic reactions by limiting the concentration of oxygen in the gas phase. He was examining how much carbon and chromium got oxidised. As expected, the carbon continued to combine with the oxygen, thereby decarburising the melt.

But, for some strange reason, the use of a mixed gas essentially stopped the oxidation of chromium. This should not have happened, because argon is inert. And Krivsky's results were not a fluke, they proved to be reproducible. But unexpected results are not uncommon in research, which by definition ventures into the unknown. It takes vision to realise their significance, beyond the ordinary method of scientific publishing. There also is some serendipity involved. Something in Krivsky's experiments attracted an intense interest from the highest echelons of Union Carbide. This is because the results were not ordinary. The loss of expensive chromium due to oxidation was a well-known cost in the manufacture of stainless steel. Krivsky's experiments eliminated chromium oxidation although that was not the original intent of his experiments. If these results could be reproduced on a larger scale, there was a potential to reshape the stainless steelmaking industry. The original mission of resolving the conflicting published data in metallurgical thermodynamics remained temporarily unresolved, replaced by an opportunity so much bigger in scope and impact.

1.3   Hope for a Big Change

The making of steel includes a process in which the carbon concentration of the liquid is controlled to the desired level by reaction with gaseous oxygen. When the melt contains chromium which is essential to render the steel stainless, it too will tend to oxidise. Traditionally, this was grudgingly accepted with the chromium concentration later adjusted to the required level by adding an alloy of iron that is rich in chromium (low-carbon ferrochrome); the entire process was implemented in the electric arc furnace responsible for melting the components in the first place. The special ferrochrome with its low-carbon concentration is about twice as expensive per kg of chromium as the commonly available, inexpensive variety which has too much carbon to be tolerated in stainless steel manufacture. If only a way could be found to use the high carbon variety of ferrochrome, the cost savings would be very significant! One hundred tons of stainless steel typically contains about 18 tons of chromium. The order of magnitude of cost savings for a 100 ton heat of stainless steel by switching from the low to the high carbon varieties of ferrochrome would be about $18,000 in the cost of chromium alone. Apart from the lower material cost, the productivity of the electric arc furnace could be enhanced if the oxygen treatment could be carried out in a separate vessel, the converter. But all these benefits were out of reach in the traditional electric furnace stainless steel making with low-carbon ferrochrome as the source of chromium. There was no known method of using low cost, high carbon variety of ferrochrome as the chromium source, because the added carbon had to be removed by oxidation and that meant that the chromium would also get oxidised. Krivsky's experiments indicated a solution although the underlying scientific principle remained a mystery. What had happened was unexpected, if reproducible on a large scale, the idea would be a no-brainer for all involved in the production of stainless steel.

1.4   From Experiment to Practice

Union Carbide was not a stainless steel producer, but there were powerful incentives to exploit Krivsky's experiments. First, there was the possibility of huge cost savings from the use of the cheaper high carbon variety of ferrochrome and achieving significantly higher productivity in a converter process that is inherently more rapid. Second, there was a growing demand for the particularly low-carbon variety of stainless steels because during welding, the carbon can cause local depletions in the chromium concentration by forming compounds. The depletion destroys the stainless character, leading to a pernicious decay in the properties. It was expensive to produce these low-carbon varieties using the conventional process because the greater temperatures necessarily degrade the refractories that line the furnaces.

One factor in favour of Krivsky's idea was the ready availability of argon, because Industrial oxygen is distilled from liquified air that contains argon, which has a different boiling temperature and therefore can be extracted in the same process. The anticipation of a breakthrough was stimulating efforts in increasing the size of experiments. There were encouraging results when tests with 45 kg steel were scaled up in a 1000 kg electric furnace, albeit with procedural difficulties such as splashing of the molten-metal bath and refractory erosion, but the results confirmed the previous work so even larger scale tests were planned with some optimism [1]. The furnaces capable of handling 3000–5000 kg, were available at the Haynes Stellite Company, a separate division of Union Carbide. The results disappointed. Argon was found to be ineffective when blown above the surface of the melt as was previously done in a 1000 kg melt. A clear set-back given the scale of the experiments, but deep down, these results could not be a final verdict of failure. It was time for a fundamental change in the process design. In fact, encouraging results were obtained when argon was injected within the bath. Full commercial trials needed even larger furnaces not available at Union Carbide so an agreement for collaboration was reached with Joslyn Stainless Steel Company, which was much smaller than Union Carbide but courageous nevertheless. The tests demonstrated that when the furnace diameter was large, the selective oxidation of carbon did not work very well and it was necessary to inject the gas mixture deep into the liquid melt [1]. It took about 7 years before the first successful full scale heat in a 15,000 kg furnace was completed in a design that allowed both oxygen and argon to be injected within the melt. Krivsky's initial experiments started in 1954 and the first successful production melt was completed in 1967. The process is now known as the argon–oxygen decarburisation process or in short, the AOD process [1].

1.5 How It Works

The AOD process is illustrated in Fig. 1.2a and an operational furnace shown [2] in Fig. 1.2b. Oxygen and argon are blown into a steelmaking reactor through a gas injection port known as a tuyère placed on the side wall near the bottom of the vessel. At the high temperature of the melt, the oxygen quickly reacts with the alloying elements in front of the tuyère forming oxides of chromium and other solutes. This oxidation would mean the permanent loss of valuable alloying elements from the liquid iron melt, unless the oxides can be reduced within the furnace away from the tuyères to recover them into solution with the molten iron. The oxides are lighter than the iron melt so they tend to move upwards with the swarms of rising gas bubbles. During this journey, most of these oxides are reduced by carbon dissolved in liquid iron. For example, the chromium oxide (Cr_2O_3) can be reduced by carbon to produce chromium metal and carbon monoxide (CO) gas by a chemical reaction shown below.

$$Cr_2O_3 + 3\underline{C} = 2\underline{Cr} + 3CO \qquad (1.1)$$

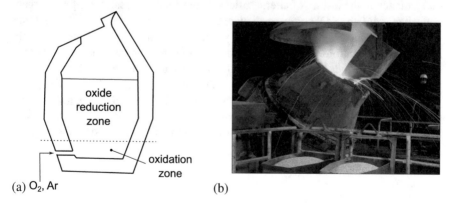

Fig. 1.2 (**a**) Schematic diagram of an AOD converter [3, 4]. (**b**) An AOD converter in action [2]

The reduction of chromium oxide by carbon dissolved in iron is seminal to the recovery of chromium which would be otherwise be lost as its oxide, Cr_2O_3. Equation (1.1) does not explicitly show any role of argon which is an inert gas. So, how does argon dilution work? A chemical reaction is facilitated when the concentrations of the reaction products are reduced. Argon when mixed with carbon monoxide does just that. It dilutes CO and hence facilitates the reduction of chromium oxide into chromium, thus preventing its loss and at the same time removing carbon from the melt as CO. When this principle is implemented in a practical stainless steelmaking process, huge cost savings result from the use of low cost, high-carbon ferrochrome.

1.6 Concluding Remarks

The original research which aimed to resolve the minor difference between the thermodynamic data reported by Richardson and the results of Hilty and Crafts was left incomplete. However, Krivsky observed: "Those differences had admirably served their purpose" [1]. "The pay backs on technical innovation in basic industries ... are larger in magnitude than anticipated." Innovations in metallurgy can affect the quality of life because of the incredible dependence we have on steels and metallic materials to function on a daily basis.

Krivsky's work in converting the spark of an unexpected result into the widespread commercial adaptation of the AOD process is revealing. The first observations were contrary to common sense, but they were not wrong. He refused to treat the results as flukes and proved them repeatedly. The leadership of Union Carbide realised the potential but also the fact that not all facilities and resources necessary for implementation could be found within. There followed the collaboration with smaller companies. Finally the development of a promising idea to a commercial process involved overcoming many difficulties, disappointments and patience in the testing of alternate designs until the successful design emerged. But when it did, the AOD process disrupted the stainless steel industry by providing a boost to productivity and metal yield while eliminating the need for the costly low-carbon ferrochrome. The stainless steel manufacturing technology had changed for the better so the AOD process is now used by an overwhelming majority of manufacturers worldwide. The economic, environmental and health benefits [5] of advanced, cost-effective stainless steel facilitated by the innovation are, like the best of technologies, not known to the general public, but they feel its consequences through the myriad of affordable products created by the stainless steel industry.

1.7 Terminology

Alloying elements	Solutes added to an element which forms the major constituent of the alloy. Carbon is an alloying element in all steels. Chromium is an essential solute in all varieties of stainless steels.
Engineered materials	Useable materials designed deliberately using engineering principles.
Ferrochrome	An alloy of iron containing a large amount of chromium. An inexpensive variety of ferrochrome often contains significant quantities of carbon.
Metallurgist	A person with a deep knowledge of the science and engineering of production, processing, structure and properties of metals and alloys.
Schematic diagram	A simplified diagram that illustrates the essence of a system, sometimes exploiting symbols, but is not an actual picture.
Thermodynamics	A subject useful for understanding how different forms of energy can be converted; it relates variables such as temperature and pressure and is able to comment on the tendency and stability of certain systems.
Tuyère	A nozzle through which gases are introduced in a furnace.

References

1. W.A. Krivsky, The Linde argon-oxygen process for stainless steel; a case study of major innovation in a basic industry. Metall. Trans. **4**, 1439–1447 (1973)
2. 2012: http://en.wikipedia.org/wiki/Argon_oxygen_decarburization
3. T. DebRoy, D.G.C. Robertson, A mathematical model for stainless steel making. Part I: argon-oxygen and argon-oxygen-steam mixtures. Ironmak. Steelmak. **5**, 198–206 (1978)
4. T. DebRoy, D.G.C. Robertson, J.C.C. Leach, A mathematical model for stainless steel making. Part II: application to AOD heats. Ironmak. Steelmak. **5**, 207–210 (1978)
5. C.P. Cutler, Economic and environmental benefits which stainless steels offer the water industry. Paper presented at fourth European Stainless Steel Science and Market Congress (2002). http://aplicainox.org/agua/wp-content/uploads/2011/11/DA02.pdf, downloaded on 27 April 2014

Chapter 2
Dazzling Diamonds Grown from Gases

2.1 What Is a Diamond?

The C–C bond is of a type where electrons are shared between the pair of atoms in such a way that it permits the formation of the giant structures, including that of diamond. Each atom contributes one electron to the shared pair and each carbon atom is able to do this with four others. These bonds are very strong, making diamond the hardest known natural material. Friedrich Moh, a crystallographer based in Graz, Austria, created in 1822 a hardness scale ranging 1–10, with diamond being 10 and Talc being 1. A hardened steel file has a Moh's hardness of approximately 7.

Diamond is a crystal that is made entirely of carbon atoms. In its chemical composition, it is therefore identical to all the other forms of carbon: Buckminsterfullerenes, graphene, graphite and nanotubes. However, it is only in diamond that each carbon atom is strongly bonded to another four to form a three-dimensional network of C–C bonds extending all the way through the visible crystal no matter what its size. The basic repeat pattern can be described in a cube, Fig. 2.1, known as a unit cell because it can be stacked indefinitely while filling space to generate a macroscopic lump that we see as a diamond.

A perfect diamond is transparent because there is no mechanism by which it can absorb light. In many materials, light can be absorbed by the promotion of the electrons within to higher energy levels, but the nature of the bonding in diamond means that there is a huge gap between the ground state and promoted state of the electrons when compared with the energies of light, X-rays and microwaves. Diamond is therefore transparent to these radiations. The same effect makes diamond an electrical insulator.

The most thermally conductive metal is silver but that is just a fifth of what diamond can achieve. This is because all the atoms in diamond are strongly coupled so a vibration of one atom extends to a large neighbourhood of atoms. These

© Springer Nature Switzerland AG 2021
T. DebRoy, H. K. D. H. Bhadeshia, *Innovations in Everyday Engineering Materials*,
https://doi.org/10.1007/978-3-030-57612-7_2

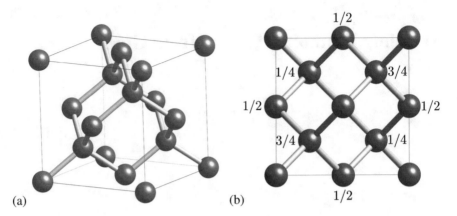

Fig. 2.1 The atomic structure of diamond. The atoms of carbon are three-dimensionally bonded. (**a**) Three-dimensional perspective of a cubic cell containing carbon atoms in the diamond arrangement. (**b**) Projection of the atomic positions on a cube face, with fractional heights of atoms in the direction normal to that face indicated. Unlabelled atoms are at heights 0,1

vibrations are therefore transmitted readily making it a powerful mechanism for the transport of heat.

A diamond in the rough has an unremarkable but nevertheless, a natural beauty associated with the pain of creation at great pressures and temperatures within the bowels of the Earth. But as a symbol of affection or an ostentatious display of wealth, it has to be cut in order to manipulate light. The process of cutting dates back to the sixth century, originating in the Indian subcontinent. Diamond, when polished, has an exceptionally large ability to reflect back any incident light, another way of saying that it has *brilliance*. The light that enters it can split into the elementary colours giving a prepared diamond the *fire*. The cutting facets the shape of the diamond in such a way that it will be reflected internally, giving the impression of *sparkle*.

Diamonds occur naturally but can also be made artificially. The first to achieve success was James Ballantyne Hannay in 1879 (published fully in 1880 [1]) in which he heated sealed-tubes of iron containing mixtures of paraffin, bone oil and lithium to red heat. We shall see next that the processes have developed dramatically over the years so artificial diamond production is now routine.

2.2 Diamonds Grown from Gases

Apart from its beguiling status as a gemstone, diamonds find uses as cutting tools, electronics, laser, microwave and X-ray transmission windows, synchrotrons, and infrared transparent domes and windows. Can naturally occurring diamonds be used in this manner? Yes, but they are not always available in the shape and size necessary

Fig. 2.2 Two diamonds grown originally from gases at Washington Diamonds Corporation by chemical vapour deposition [2]. It is claimed that these diamonds are chemically, physically and optically indistinguishable from those that are mined [3]. The company now produces diamonds weighing between 0.5 and 2.5 carats (0.1 = 0.5 g) [3]

for an application such as a large infrared transparent dome. Natural diamonds are created deep within the Earth at high pressures and when those conditions revert to ambient, the diamond is not stable when compared with graphite, for example. But luckily there is a barrier to the change from diamond to graphite, which can be triggered by heat or mechanical stimuli. The reverse can be achieved, by appropriately pressurising graphite while heated (50,000 atmospheres, 1400 °C), to convert it into an artificial diamond. This opens up the possibility of customising the shape and size. An alternative method, that of chemical vapour deposition (CVD) can create polycrystalline or single crystal diamonds (Fig. 2.2).

The CVD process is ideal for applications where a diamond in the form of a thin film suffices. The process deposits films of diamond on a suitable substrate from hydrogen-methane mixtures at pressures of about 0.04 atmospheres. To generate single crystal films requires a diamond substrate on which the film deposits epitaxially, i.e., with the crystal orientation of the film being identical to that of the substrate. In 2013, a company named *Type IIa diamond* in Singapore produced high quality single crystal diamonds using a microwave plasma CVD process starting with a diamond seed. It is now possible to make fairly large gemstone-quality diamonds for the jewellery market, but the attitude that natural diamonds, so difficult to mine, are special prevails in the uneducated mind.

Since the shape and size of the diamonds made industrially can be controlled, they have an advantage over natural diamonds for many industrial applications. However, the process is not rapid because the carbon necessary for the synthesis of diamond in the CVD process is obtained from a source gas such as methane. Making solid diamond from a much less-dense gas takes time.

The ability to synthesise diamonds of desired geometry on demand, sometimes over a large area, is exciting because it enables the fabrication of unique devices that were not possible before. The films can be patterned and configured into different three-dimensional shapes by using dry etching and masking methods (Figure 2.3). The films can be cheap; 10 × 10 × 0.5 mm film for just 50 USD. Because they

Fig. 2.3 A diamond film produced using microwave plasma-assisted CVD from CH_4-H_2 gas mixtures. There are many crystals of diamond that constitute this film. Reproduced from Wild et al. [4], with permission of Elsevier

are biologically inert, they can be used as re-useable plates for cell cultures or as coatings for implants in the context of bone and neural tissues [5].

2.3   How Is CVD Diamond Made?

In this process, diamond is deposited from a gas onto a solid substrate. The gas consists of a mixture of a hydrocarbon and hydrogen, typically 1% methane (CH_4) and 99% hydrogen (H_2). The gas is exposed to either a hot filament of a refractory metal wire which is heated to over 2000 °C or to microwave radiation. These techniques split the molecules of the gas into components such as CH_3, C_2H_2 and atomic hydrogen which facilitate diamond deposition. Diamond films have been produced on a variety of substrates such as silicon, molybdenum, tungsten and tungsten carbide. In most cases, polishing the substrate with diamond powder facilitates diamond nucleation, generating typically about 100 million tiny diamond nuclei per square centimetre [6]. The CVD process commonly produces polycrystalline diamond. This is pure but appears unattractive compared to gemstone-quality single crystal diamond. The growth of a single crystal diamond requires a similar diamond substrate on which it can grow epitaxially.

Between the hot filament and microwave deposition techniques, the former is simpler and less expensive. The important variables include the inlet gas composition, filament temperature or microwave radiation intensity, substrate temperature, pressure in the reactor and the filament to substrate distance.

A hot filament laboratory reactor used to deposit diamond on a silicon wafer is shown in Fig. 2.4. Typically, the inlet gas enters at the lower part of the cylindrical

Fig. 2.4 A laboratory reactor [2] for the chemical vapour deposition of diamond. The filament is shown directly below the substrate

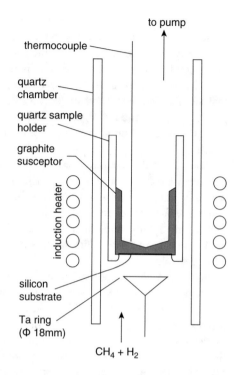

chamber and the exhaust gas goes out from the top. The diamond is grown on a silicon wafer substrate, which is kept in place by a specimen holder. The chamber contains a tantalum ring filament which is heated inductively and placed about 1 cm below the substrate. The filament and the substrate temperatures are kept at 2250 °C and 900 °C, respectively.

The exact chemical pathway for the formation of diamond is not fully understood although the importance of species such as CH_3 and H is well recognised. In the hot filament reactor, these species are generated near the filament and transported to the growth surface by convection, i.e., carried with the gases as they flow in the reactor, and by diffusion which results from the spatial gradients of concentration and temperature.

2.4 Why Is Hydrogen Needed?

In both hot filament and microwave assisted CVD of diamond, the inlet gas contains some 99% of hydrogen. When molecular hydrogen (H_2) is exposed to a hot filament or microwave radiation, a portion of it breaks down into individual atoms of hydrogen. This nascent hydrogen is far more reactive than the molecular form. First, it reacts with methane to form various hydrocarbon species such as CH_3 and C_2H_2

which also are reactive. Second, some non-diamond forms of carbon deposit in those conditions. Since atomic hydrogen reacts much faster with graphite than diamond, it selectively etches or removes graphite and other non-diamond carbon in preference to diamond, thereby eliminating non-diamond contaminants. Graphite and diamond have the same composition but different crystal structures and bonding. It is possible that atomic hydrogen prevents the transformation of a diamond surface to that of graphite [6]. Atomic hydrogen can induce new growth sites by removing hydrogen bonded to the carbon atoms on the growth surface where new carbon atoms can attach.

Apart from the chemical activity, the atomic hydrogen facilitates heat transfer from the filament to the substrate [7, 8]. This helps control the substrate temperature which is important to avoid the deposition of carbon that is not diamond and to determine the growth rates of diamond films. Since the dissociation of molecular into atomic hydrogen requires heat, the split occurs near the hot filament. When the atomic hydrogen reaches the specimen surface, its recombination into molecular hydrogen releases heat.

Figure 2.5 shows the variation of temperature with distance from a tantalum filament in helium, hydrogen and 1% methane 99% hydrogen mixture. At any specified distance from the filament, the temperature in helium was significantly lower than in either pure hydrogen or hydrogen-methane mixture. Since helium is monatomic and does not recombine, it cannot provide heat by forming molecules like hydrogen. The presence of methane in hydrogen diminishes the ability of hydrogen to transfer heat because some of the atomic hydrogen reacts with methane to form various hydrocarbon species, thus reducing its concentration.

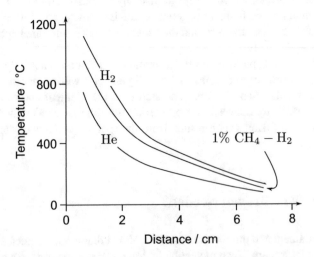

Fig. 2.5 Variation of temperature with distance from a hot tantalum filament kept at 2473 K and 0.04 atm pressure in various gas environments. Hydrogen molecules split near the filament and the recombination of atomic hydrogen on the temperature probe heats up the probe. Adapted from [7, 8]

When mechanisms of heat transfer are considered, conduction, convection and radiation normally come to mind. There is an additional mechanism, a chemical dissociation of H_2 into hydrogen atoms in the vicinity of the filament followed by their recombination at the growth surface, contributing to the heat transfer process.

2.5   Faster Rate of Deposition → Cheaper Diamonds

A greater growth rate would obviously be advantageous in the cost-effective production of synthetic diamond. The deposition rate depends on the transport of atomic hydrogen and hydrocarbon species from near the filament where they are generated, to the substrate where diamond is deposited. Convection and diffusion are the two mechanisms of transport of these species. Convection is the transport of a species because of gas flow in the reactor. The transport additionally occurs by diffusion driven by concentration gradients in the reactor. To understand the roles of these transport mechanisms, in the context of diamond growth, consider the configurations illustrated in Fig. 2.6, where both the direction of gas flow within the reactor and the relative positions of the filament and the substrate are varied.

In configuration 1, the transport of active species towards the substrate is entirely due to externally driven forced flow, in configuration 2, natural convection and forced flow both contribute to the transport towards the substrate. In configuration 3, all flows are directed away from the substrate, whereas in configuration 4, the natural convection is the only flow towards the substrate. Diffusive transport, not illustrated, occurs in all cases, driven by the concentration gradients and not affected by either the direction of flow or the positions of the filament and the specimen.

It turned out that the diamond growth rates were not very different in the cases illustrated [2] as shown in Fig. 2.7. In order to understand this, consider configurations 2 and 3, where both the forced and natural convections carry species

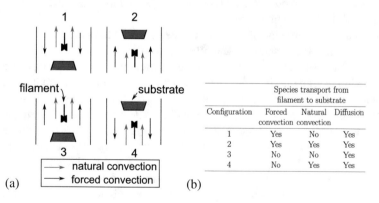

Configuration	Species transport from filament to substrate		
	Forced convection	Natural convection	Diffusion
1	Yes	No	Yes
2	Yes	Yes	Yes
3	No	No	Yes
4	No	Yes	Yes

Fig. 2.6 Experimental arrangements showing various gas flow directions and the positioning of hot filament and sample

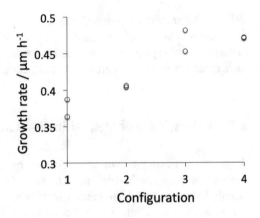

Fig. 2.7 Growth rates of diamond for various flow configurations

in the same direction. But in configuration 2, they take species toward the growth surface, whereas in configuration 3, they take the species away from the growth surface. If natural and forced convections were important, the rates would be significantly different. But the results in Fig. 2.7 show that they were not! So, convection can be judged to play a minor role in determining the rate of diamond deposition [2]. Results of configurations 1 and 4 also reinforce this finding. Here, the forced and natural convections carry species in the opposite directions. One takes the nutrient species to the growth surface and the other takes them away from the growth surface. Figure 2.7 shows that the rates were not significantly different in the two cases. Clearly, the directions of natural and forced convections did not affect the diamond growth rate.

Finally, let us consider the result for configuration 3, where both forced as well free convection carry species away from the substrate. Therefore, if convection was an important mode of species transport to the growth surface, the rate of deposition would be very small if not negligible. However, the diamond growth rate for configuration 3 is comparable to the other arrangements. Thus, the results prove that neither forced nor natural convection is an important mechanism of mass transport in the hot filament diamond growth.

Given this, the diffusion of species must be the controlling mechanism of species transport within the reactor, essentially independent of the flow configurations. The diffusion coefficient of a gaseous component is inversely proportional to total pressure in the reactor. Since the chemical vapour deposition of diamond is carried out at a very low pressure, 0.04 atm, the diffusion coefficients of various species are about 25 times greater than their corresponding values at one atmospheric pressure.

Can this mechanistic knowledge be exploited to enhance the diamond growth rate? The diffusion rate per unit area depends on the spatial gradient of nutrient species concentrations and the diffusion coefficient. The diffusion coefficient can be increased by reducing pressure, but there is a compromise to be achieved because the concentrations of various nutrient species in the gas will decrease at the same time. The balance needs to be achieved through trial experiments, but the mechanistic knowledge could influence the design of CVD diamond reactors. For example, given that the flow of gases through the reactor plays the minor role in active-species transfer, the locations of the gas inlet and outlet that determine the flow pattern of gases through the reactor may not be important. The design engineers therefore have significant flexibility in the fabrication of such reactors [9].

The experimental results of the rates of deposition of the diamond films in Fig. 2.7 can serve as an order of magnitude indicator of the time necessary to grow a chunk of diamond. Figure 2.7 shows that the diamond growth rate was roughly about $0.5 \, \mu m \, h^{-1}$. With optimisation, the growth rate might be increased; a doubling of the rate over an area of $3 \, cm^2$ to $1 \, \mu m \, h^{-1}$ would still require about 8 days to grow 1 caret (200 mg) of diamond. The basic reason for this pace is that the density of the solids is much greater than that of gases, so the process necessarily takes time. But the reward would be a single crystal of diamond that dazzles not only in its perceived beauty, but excels in its properties.

2.6 Epilogue

Many industries now synthesise diamond so the artificial form is commercially available with many combinations of characteristics that compete against natural diamonds, including carat, cut, clarity and colour. The colour is achieved by doping the diamond with other atoms or through structural defects. The ability to synthesise such an amazing material is transformative. A diamond heat sink can carry heat many times faster than copper, a diamond window can transmit infrared radiation with so little loss that it can serve as a premium durable laser window, and a wide band gap diamond semiconductor can operate at temperatures much higher than the current generation of semiconductor materials. Diamond-tipped tools are ideal for cutting hard materials and diamond grit incorporated into cutting wheels makes the task so much easier.

Is the man-made diamond going to enjoy the same prestige as the natural diamond? Not as a gemstone because of the irrational yearning for the natural version, a yearning that has been cultivated by the mining industries over the centuries. This continues to be the case using clever advertising and internal marking of the natural diamonds with lasers, so that the client can be sure it is obtained by extremely tedious digging of the earth to many kilometres of depth with miners working in harsh environments. How does this make sense? Nevertheless, the industrial diamond is serving many more useful needs that silently serve society— they do not need to dazzle to function!

2.7   Terminology

Band gap	It is the energy needed to excite an electron from bound state to free state.
Chemical vapour deposition	A method to produce thin films of a solid material from gases at controlled temperature and pressure.
Covalent bonds	When atoms of different elements share electron pairs, the electron shells are filled up and the atoms gain stability. Covalent bonds can exist in gases, liquids and solids.
Epitaxial	Atomic arrangements of a new growth layer conforming to the structure of the existing layer.
Face-centred cubic	Atomic arrangement in space which is repeated in a manner that the atoms occupy corners of the cube and the centre of its faces.
Induction heating	A form of non-contact electrical heating of electrically conducting materials. An alternating current is applied through a coil that surrounds but does not touch the part to be heated. A circulating current called the eddy current is established within the part which gets heated due to resistive heating.
Infrared	Radiations of wavelengths between 800 nm and 1 mm emitted from hot surfaces. These wavelengths are higher than that of the red light but lower than that of the microwave radiation.
Laser	Acronym for *light amplification by stimulated emission of radiation.* It is a powerful device that emits a band of light which can stay narrow over a long distance. Because of its high energy, it is used to cut and weld metals and alloys.
Nuclei	Plural of nucleus. Nucleus is the core on which the new phase grows.
Refractory metals	Metals such as tungsten, tantalum, molybdenum, rhenium and niobium can function reasonably well at very high temperatures because they have high melting points, good high temperature strength and high resistance to wear and corrosion.
Susceptor	Induction heating cannot be used directly to heat an electrically non-conducting material. This problem is often overcome by induction heating of an electrically conducting material called a susceptor which then transfers heat to a non-conducting work piece by conduction, convection and radiation.
Synchrotron	A machine to accelerate charged particles to very high energies in a circular path by an electric field to generate X-rays and other radiations.
Thermocouple	When two dissimilar alloy wires are joined at one end and the joint is heated, a voltage develops between the colder portion of the two wires. This voltage is a measure of temperature of the junction.

References

1. J.R. Hannay, On the artificial formation of diamond. Proc. R. Soc. London **30**, 200–205 (1880)
2. T. DebRoy, K. Tankala, W. Yarbrough, R. Messier, Role of heat transfer and fluid flow in the chemical vapor deposition of diamond. J. Appl. Phys. **68**, 2424–2432 (1990)
3. Washington diamonds corporation, Wikipedia article (2014). http://en.wikipedia.org/wiki/ Washington_Diamonds_Corporation. Accessed 13th Nov 2014
4. C. Wild, R. Kohl, N. Herres, W. Müller-Sebert, P. Koidl, Oriented CVD diamond films: twin formation, structure and morphology. Diam. Relat. Mater. **3**, 373–381 (1994)
5. P.A. Nistor, P.W. May, Diamond thin films: giving biomedical applications a new shine. Interface **14**, 20170382 (2017)
6. W. Yarbrough, R. Messier, Current issues and problems in the chemical vapor deposition of diamond. Science **247**, 688–696 (1990)
7. K. Tankala, T. DebRoy, Modeling of the role of atomic hydrogen in heat transfer during hot filament assisted deposition of diamond. J. Appl. Phys. **72**, 712–718 (1992)
8. Y.A. Yarbrough, K. Tankala, M. Macray, T. DebRoy, Hydrogen assisted heat transfer during diamond growth using carbon and tantalum filaments. Appl. Phys. Lett. **60**, 2068–2070 (1992)
9. K. Tankala, T. DebRoy, Transport phenomena in the scale-up of hot filament assisted chemical vapor deposition of diamond. Surf. Coat. Technol. **62**, 349–355 (1993)

Chapter 3
Stirring Solid Metals to Form Sound Welds

3.1 Introduction

The most common process of joining metals involves localised melting of the pieces to be joined by application of heat; the molten metal solidifies to make an integral joint. This *fusion welding* is incredibly successful so much so, that almost all engineering structures such as bridges, buildings, ships, pipelines and vehicles are put together in this way. However, the local intense heating, the melting and complex cooling can leave the assembly in a state of undesirable permanent stress. It may also cause distortion, i.e., an unintended change in the shape of the assembly, which can accumulate over large distances to render the object unviable in its intended application. All of these issues can, and are managed effectively so fusion welding is here to stay, but a new process, designated *friction stir welding* (FSW) opened up innovative approaches to joining, particularly of the softer metals (Fig. 3.1).

Invented by Dr Wayne Thomas at The Welding Institute, UK, in 1991 [1], FSW is in essence a solid-state process so the peak temperatures reached are lower than the melting temperature of the component. It avoids completely any issues associated with melting and solidification. It does not require filler metals or fluxes but achieves the joint by the application of local, severe plastic deformation. This limits, though it does not eliminate, distortion of the final product. The equipment used is in principle simple, though powerful and precise. A hard, rotating tool with a pin that digs between the plates to be joined, and a shoulder that generates heat by friction are translated along the joint line, Fig. 3.2. Near the tool, the alloy is softened by heat. The tool stirs the softened material to help its transport from the front of the moving tool to its trailing end where the joining occurs. Rigid fixtures hold the plates in place under pressure so that the hot softened alloys are forged into a strong joint.

The world's largest robotic FSW machine, about 52 m tall and 24 m wide, is located at NASA's Michoud Vertical Assembly Center in New Orleans. It is now used to construct the world's most powerful rocket, a part of NASA's Space Launch System. This facility builds both the core stage of the rocket that includes

© Springer Nature Switzerland AG 2021
T. DebRoy, H. K. D. H. Bhadeshia, *Innovations in Everyday Engineering Materials*,
https://doi.org/10.1007/978-3-030-57612-7_3

Fig. 3.1 Friction stir welding is a radically innovative solid-state welding technique that involves heating and softening of alloys, but no bulk melting. Aluminium alloys and a few other soft alloys are now commercially welded by FSW. Many other materials have been tried in laboratories. Photo credit: Dr. Chobi DebRoy and Dr. Tuhin Mukherjee

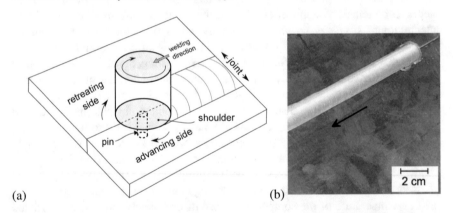

(a)   (b)

Fig. 3.2 (**a**) Principle of friction stir welding. The plates are held rigidly while the rotating tool is translated along the welding direction. Other joint configurations in including the joining of cylindrical objects are also possible. (**b**) An actual friction stir weld between two aluminium plates. The arrow indicates the welding direction

domes, rings and barrels and measures 65 m tall and 8.4 m in diameter and contains cryogenic tanks to store liquid hydrogen and liquid oxygen to feed four rocket engines. About 2.5 km of FSW is necessary for the core stage of the launch system. Figure 3.3 shows a 40 m tall cryogenic fuel tank made by friction stir welding of an aluminium alloy. It can store 537,000 gallons of liquid hydrogen.

FSW is now a process of choice for aluminium and other soft alloys; this particularly is the case for simple geometries because of the need to translate the tool by mechanical means. There are potential applications to other metallic systems. Because of its many advantages, it has found widespread use in aerospace,

Fig. 3.3 The core stage liquid hydrogen tank, about 130 feet tall, friction stir welded at the Vertical Assembly Center at NASA's Michoud Assembly Facility. The tank holds 537,000 gallons of liquid hydrogen [2]

shipbuilding, automotive, railway and electronic industries. Smooth operation of friction stir welding involves complex interactions between a variety of simultaneous physical processes such as heat transfer and flow of softened metal around the tool. The flow of heat is aided by the flow of material around the tool and it affects the heating and cooling rates and the microstructure and properties of the welded joints.

3.2 Only so Hot, Not Hotter

The alloy needs to be softened by heat to flow around the tool and forge into a sound weld. Temperatures needed for softening depend on the specific alloy as shown in Fig. 3.4a. Steels are harder than the aluminium alloys and higher temperatures are needed to soften steel. So, more heat needs to be generated for the FSW of steels than for aluminium alloys. The peak temperature is a good measure of the heat generation rate because it depends on the amount of heat generated per unit length. A large contact area between the tool and the material and a high tool rotational speed both result in a high rate of heat generation. In contrast, a high welding speed yields a low rate of heat generation per unit length of the work piece. The peak temperature in its non-dimensional form, T^*, can be estimated [3] from the dimensionless heat input, Q^* (Fig. 3.4b). Here, the non-dimensional peak temperature is the ratio of peak temperature to the solidus temperature. The dimensionless heat input, Q^*, depends on the rotational and translational speeds of the tool, tool dimensions and the properties of the alloy and the tool materials.

Temperatures estimated from Fig. 3.4b are reliable for the range of Q^* illustrated; the computed peak temperatures are within 3–9% of the corresponding experimentally determined values. The peak temperatures depend on the alloy. For example, in both the experimental results and estimated values, the peak temperature is for the

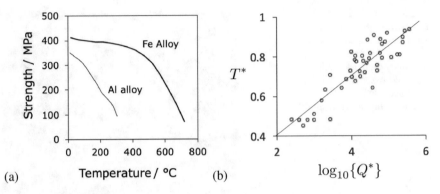

(a)   (b)

Fig. 3.4 (**a**) Steels retain their strength up to a much higher temperature than aluminium alloys. So FSW of steels requires higher temperatures than aluminium alloys [4]. (**b**) Dimensionless temperature, T^*, versus log of dimensionless heat input, Q^*, for a large variety of metallic alloys [3]

same welding conditions. The peak temperatures in FSW are between 70 to 85% of the solidus temperatures of the alloys. Since the highest temperatures attained during FSW is less than the solidus temperature of the alloy, there is no danger of bulk melting within reasonable limits of welding variables.

3.3 Race to Join

The bits of material affected by the severe deformation associated with the motion of the tool have to slide in such a way that a defect-free joint is produced in a time frame that ensures high productivity. It is important then to be aware of the vector field defining the complex motions involved because these also help to calculate the forces experienced by the tool.

Most of the heat is generated at the tool surface by its friction with the work piece. As a result, high temperatures are reached near the shoulder. The heating softens the metal, enabling the smooth flow of the material around the tool to form the joint. The flow, however, is not uniform (Fig. 3.2). The rotation of the tool draws the material from the front to its trailing end on one plate and in the opposite direction in the other plate. As a result, on one side of the tool called the advancing side, the direction of motion of the softened material is the same as the direction of welding. On the other side of the tool called the retreating side, the softened material flows in a direction opposite to the translation of the tool along the welding direction. Therefore, the two sides of the weld differ in the nature of flow, heat transfer and the structure and properties of the joint.

Three features contribute to the character of the flow necessary to ensure structurally sound joints. First, a slug of plasticised material rotates around the tool, driven by the friction between the tool and the work piece. Second, close to the pin,

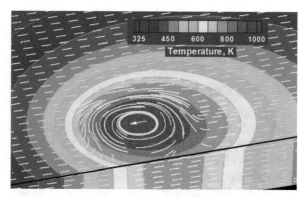

Fig. 3.5 A typical computed material flow path with respect to the tool. The asymmetric nature of the flow is apparent

the threaded pin pushes the material downwards. This push drives an equal amount of material upward a short distance away from the pin. Finally, the translation of the tool along the welding direction helps the movement of the material to the back end where a weld is formed from the interaction of all these effects.

The flow during FSW has been studied experimentally. Tracer materials within the work piece with their positions defined before and after welding indicate the trajectories and this can include time lapse snapshots from selected locations though the methods do not enable the entire velocity field to be determined. The flow stress of the material can be characterised in simpler experiments as a function of temperature, strain and strain rate with the resulting data used in sets of equations representing simultaneously the conservation of mass, momentum and energy. Their solution provides detailed temperature and velocity fields in the entire flow domain alongside other parameters such as the torque and traverse force on the tool, all as a function of time.

The velocity of the softened material depends on the welding parameters and the specific tool geometry used. Velocities are roughly an order of magnitude smaller than those of the tool shoulder above the location [4]. The velocities and the paths that the metal located in the deformed region travels to form the weld can be simulated in a computer. Figure 3.5 shows an example of the calculated motion of the material with respect to the tool. The width of the path for material flow near the pin depends on the local temperature, tool design and welding parameters such as rotational and welding speeds. They also affect the entire flow pattern of the softened alloy. Because of the interaction between the translational and rotational components of the velocity, the flow of the alloy is not symmetric around the tool. In fact, more materials flow along the retreating side than the advancing side.

A streamlined flow of material is essential during friction stir welding. Inadequate transport that breaks continuity along the advancing side causes a defect known as a *wormhole* which essentially is pore that is extended to comply with the overall flow pattern. The flow pattern also affects the torque and traverse force which together with the properties of the material and the tool affect the durability of the tool. Tool durability has a major impact on the commercial feasibility of FSW, especially for hard alloys such as steels.

3.4   The Durable Stirrer

Tools drive the process of welding because they affect the generation of heat which facilitates the plastic flow of the softened alloys. The interaction between the tool and the work piece also leads to a gradual deterioration by wear of the tool geometry. Consequently, the durability of the tool critically affects the cost and practicality of welding. Tools typically have a round shoulder and a threaded cylindrical pin, although many geometric variants are used (Fig. 3.6). In excess of 40 different types of welding tool have been developed for thin walled structures, heavy and thick walled structures. Retracting pin FSW tooling allows the joint to be prepared without showing any signs of the penetration of the tool into the work piece when the welding process is terminated.

The strength, fracture toughness, hardness, thermal conductivity and thermal expansion coefficient of the tool affect its wear and therefore the weld quality and performance. An alloy containing hard particles will require a tool material that resists abrasion. Since the work piece is in close contact with the tool, the eroded tool materials are embedded in the weld microstructure and may detrimentally affect its properties. Excessive wear increases the cost of FSW. Tool materials must also resist oxidation at elevated temperatures and ideally should not mix with the work piece material. A well-designed tool must be capable of resisting erosion and fracture leading to premature failure, especially for the FSW of hard alloys such as steels.

Tools experience a harsh environment of severe stress and high temperatures for extended time. Rotation at high speed and linear motion of tool through the softened but solid alloy make them susceptible to wear. Tools may also deform plastically due to a reduction in their yield strength at elevated temperatures during FSW. Stresses on the tools are important because tools fail when exposed to stresses higher than their ability to endure them. These stresses depend on the strength of the work piece at high temperatures encountered during FSW. Temperatures depend on the welding parameters and the properties of tool such as its geometry and physical properties. The coefficient of thermal expansion also affects the thermal stresses in the tool. So, tool durability depends on its design, properties of the tool material and work piece, and the welding conditions. Tool durability is often limited by the strength of the work piece and to a lesser extent, the welding variables.

Cost-effective and durable tools are available for the FSW of soft alloys such as aluminium or magnesium alloys. These alloys and aluminium matrix composites are usually welded using steel tools. In addition, steel tools are also used for joining dissimilar alloys in both lap and butt configurations. Availability of cost-effective and durable tools has been a critical factor for the widespread commercial success of FSW for the welding of soft alloys.

Fig. 3.6 Friction stir welding tools. (**a**) Cylindrical threaded-tool made of steel for welding aluminium or magnesium alloys. (**b**) Silicon carbide tool for welding 10 mm thick steel. (**c**) Polycrystalline boron nitride tool for welding 15 mm thick steel. (**d**) Tungsten carbide tool. Pictures courtesy of Hidetoshi Fujii

Boron nitride (BN) is a preferred tool material for hard alloys such as steels and titanium alloys because of its high strength, hardness and stability at elevated temperatures. Its low coefficient of friction results in a smooth weld surface. However, the high temperatures and pressures required for its manufacture make the tool very expensive. It also tends to fail during the initial plunge stage because of its poor fracture toughness. The maximum depth of a FSW weld is limited to about a centimetre for the welding of steels and Ti alloys using BN tools and the length of an acceptable weld will be determined by how long the tool survives. Polycrystalline boron nitride performs well with steel and iridium is the ultimate though expensive choice as a tool material. Tools made of tungsten alloys, although not as hard and wear resistant as boron nitride, are more affordable options. Commercially pure tungsten is strong at elevated temperatures but has poor toughness at ambient temperature and wears rapidly when used as a tool material for these alloys. The properties of tungsten can be improved by adding rhenium, but the cost then increases significantly. The high cost and short life of tools is a limiting factor in the application of FSW to iron, nickel and titanium alloys.

3.5 Structure Inside the Weld

The weld is naturally not uniform and there are features inside that affect the properties. The weld is divided into four regions as shown in Fig. 3.7. Base metal is the original material unaffected by welding. The region just outside the shoulder is the heat affected zone where the temperature is high enough to affect some features of the structure in micrometre scale. These changes can be seen under a microscope. Beneath the shoulder of the tool, the solid alloy is vigorously stirred and this region is identified as the thermomechanically affected region (TMAZ). The nugget represents the severely stirred metal which induces a fine, equiaxed grain structure, finer than any of the surrounding regions.

FSW is most widely used for the welding of lightweight, strong and corrosion resistant aluminium alloys. They typically contain magnesium, zinc, copper, silicon and other alloying elements to enhance properties and are classified into heat treatable and non-heat treatable alloys. The FSW variables and the properties of the work piece both affect the thermal cycles, strains and strain rates, and the structure and properties of the welds.

The heat treatable aluminium alloys gain their strength from fine dispersions of precipitates, which are small crystals that form inside the metal and obstruct deformation. The heat from welding can cause them to coarsen, rather like soap bubbles, resulting in the metal losing some of its strength. They may even dissolve in the region close to the tool where temperatures reach peak values. During subsequent cooling, some re-precipitation may occur. It is common for heat treatable aluminium alloys to have a reduced hardness in the heat affected zone because of competition between the dissolution and re-precipitation. The change in the

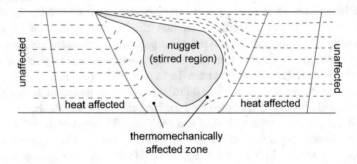

Fig. 3.7 Schematic illustration of the heterogeneous structure created within and in the vicinity of the weld. Adapted from Frigaard et al. [5]

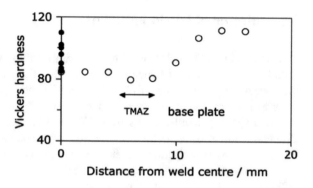

Fig. 3.8 Open circles are data from friction stir weld, plotted relative to the centreline of the FSW for an alloy designated AA2219. Filled circles from a variety of arc and electron beam welds

precipitate structure is augmented by the deformation of the work piece during FSW which introduces numerous dislocations. These changes in the grain structure also affect the overall weld properties.

One of the common aluminium alloys contains about 6.3 wt% Cu; the copper can be induced to precipitate a compound Al_2Cu which increases its strength and the intrinsic toughness of aluminium at low temperatures. The alloy frequently is used as a container material for liquified gases after cold working, solution treatment and artificial ageing. Arc welding processes result in a reduction in the strength depending on the strength of the thermal cycles. During FSW, the Al_2Cu precipitates coarsen within the TMAZ and deformed crystal recrystallisation results in fine grains in the harder fusion zone. The coarsening of precipitates leads to softening and is detrimental to the microstructure. FSW does not seem to have an advantage over fusion welding with respect to the strength of the weld zone as shown in Fig. 3.8. It is not surprising that the precipitation hardened alloys become weaker after FSW, since the majority of strengthening in most strong alloys originates from the presence of precipitates.

Non-heat treatable aluminium alloys derive their strength from cold work and solid solution strengthening. In contrast to precipitation-strengthened alloys, where the hardness decreases in the HAZ, FSW of non-hardenable AA 5082 results in rather uniform hardness across the weld. This general scenario may be complicated by the effects of deformation during FSW where the strain and strain rate affect the structure of the welds.

Welds of non-heat treatable aluminium alloys also show a thermomechanically affected zone and a recrystallised weld nugget. The heat from the FSW results in annealing and recovery of the work hardened structure. As a result, the fine-grained recrystallised nugget becomes softer than the base metal. The grain size increases with the increase in peak temperature which is often caused by greater tool rotation speed. These complexities clearly indicate the need for detailed characterisation of structure and properties of welds for the development of any new FSW application.

3.6   Stirring of Dissimilar Alloys

FSW is very attractive for the welding of dissimilar alloys which differ in properties, chemical composition or structure. The fact that there is no bulk melting makes the process especially attractive where melting adversely affects the structure or properties of welds. So there has been considerable interest worldwide for the welding of dissimilar alloys by FSW. Familiar examples include welding of aluminium alloys to copper for electrical and electronic industries, to titanium alloys for aerospace and transportation industries, and to steels for the automotive industry.

There are three main issues in the FSW of dissimilar alloys. They are the differences in the deformation behaviour of the alloys, formation of detrimental intermetallic compounds and the differences in physical properties such as the thermal conductivity of the two alloys. They contribute to the asymmetry in both heat generation and material flow during FSW and affect the structure and properties of the welds.

Engineers are testing several potential solutions to address these difficulties. For example, since the flow behaviour of the two dissimilar alloys may be different, it is necessary to choose which alloy should be placed on the advancing or retreating sides. This is important since the outcome of the FSW of dissimilar alloys is affected by the asymmetry in temperature and stress between the advancing and retreating sides. The extent of asymmetry depends on the properties of the two alloys and the welding parameters. Since most material flow takes place on the retreating side, the softer material needs to be placed in the retreating side. It is also important to select the exact location of the tool relative to the original joint. Keeping the tool pin within the soft material prevents rapid tool wear and increases its life. Prevention of tool wear is particularly important for the welding of alloys that retain their hardness at elevated temperatures.

Joining of aluminium alloys with steels is an important need in many industries. Preventing a direct contact between the FSW tool and the steel is key to avoiding rapid tool wear. An interesting arrangement is to place a steel tool pin inside the aluminium part, a short distance away from the steel–aluminium interface. This arrangement works well in a laboratory. In principle, aluminium can be welded to steel with the ordinary tool avoiding any direct contact with the steel part except at the shoulder. The direction of tool rotation needs to be adjusted so that the aluminium alloy is in the retreating side of the joint, since the softer alloy flows easily. In addition, the higher temperatures on the advancing side help to soften the harder steel. The welding parameters such as the rotational and welding speeds, tool design and other variables are to be optimised to obtain sound welds.

In all cases, the position of the tool with respect to the original joint interface affects the strength and ductility of the joints. So, it is important to optimise the tool offset distance. Again, since the tool rotation direction should be selected such that the softer material is in the retreating side, both the tool rotational and the welding speeds are important variables. They both need to be fine-tuned for each alloy combination. Welding of aluminium alloys with aluminium metal matrix composites is another unusual application where traditional fusion welding processes are not appropriate and FSW provides a unique opportunity.

In fusion welding, the alloying elements mix with the consumables on an atomic scale because of a strong recirculating liquid metal flow in the weld pool. In contrast, during FSW, the mechanical alloying does not lead to perfect mixing and true solid solutions do not form in bulk scale. For example, when joining copper with an aluminium alloy, the welds often contain isolated copper regions within the aluminium side of the nugget due to imperfect mixing. There is also an optimum welding speed to allow adequate mixing of the two alloys. Both lower and higher welding speeds produce weld defects. At low welding speeds, there is enough time for the mixing of copper and aluminium but the exact mechanism for the defect formation is not known. At high welding speeds, there is incomplete mixing of the alloys leading to defects near the weld interface. So, the welding variables need to be carefully controlled to avoid defect formation.

3.7 Expanding Its Reach

The success of FSW in the welding of aluminium alloys has stimulated exploration of its applicability to other alloys such as steels, titanium, nickel and copper alloys. However, steels represent by far the greatest opportunity because of their undisputed prominence in structural applications owing to their high strength, versatility and low cost. These very factors make it difficult to apply FSW to steels. The steel must be sufficiently softened to make a sound weld. Figure 3.4 shows that an FSW tool would have to endure much harsher environment for steel compared to aluminium alloys unless the steel is heated to more than 800 °C. Cost-effective tool materials which survive such conditions for extended service remain to be developed.

Because of the high cost of tools and the availability of cheaper and more effective methods for welding steels, it remains doubtful that FSW can replace fusion welding at least in the near future. The cost of making spot welds in automobiles is a few cents of a dollar. FSW needs to compete at this cost level. During FSW, steel parts become red hot and austenite forms during the heating cycle. Austenite then transforms to various phases while cooling. The composition, structure and morphology of these phases determine the properties of the welds. The metallurgical transformations expected from cooling rates alone are unlikely to be significantly different from fusion welds. However, because the peak temperatures achieved are lower than in fusion welding, the austenite grain structure of the heat-affected zone is expected to be finer. This would be beneficial in avoiding transformation to hard, detrimental phases especially for high carbon steels.

With few exceptions, only elementary mechanical properties of steel welds have been characterised; most reports are limited to simple bend, tensile and hardness tests. For serious structural applications, it would be necessary to assess fracture toughness and other complex properties in much greater depth. Certainly, the early optimism that FSW will become a commercially attractive method for the FSW of steel in pipelines, ships, trucks and railway wagons cannot come to fruition until the welds are adequately evaluated.

The fundamental problem is two-fold, the lack of durability of the tool material and its high cost. Tools must be reused for extended times for them to be economical. This is not the case with any of the tool materials available today for the FSW of steels. It is possible that a more sophisticated hybrid welding technique involving FSW, in which a different heat source provides additional heating, may help reduce the demands on the tool material. However, anything that adds complexity is also likely to add costs. The development of a reliable, lasting and cost-effective tool material is an important unsolved problem now. So, this is where the potential benefits in research could be large. The focus must be on cost if real success is to be achieved, although this would be mitigated by the identification of critical problems which cannot be addressed by fusion welding.

A starting point for useful research may be a niche problem dealing with expensive components to justify costs. One example which admittedly would need detailed analysis is the joining of mechanically alloyed yttria dispersion strengthened iron-based alloys. These alloys are currently being investigated for the fusion research program, and there is no efficient joining technology available for them now. These are expensive materials for a critically important application. Another clear example is the underwater joining of steels where FSW would have clear advantages over fusion welding. Underwater pipelines are extremely expensive to place in position and it is possible that the cost of tooling might then become tolerable. Development of these applications would require considerable resources to better understand both the process and the resulting structure and properties of welds.

3.8   Concluding Remarks

FSW has been a remarkable achievement that has benefited many commercial applications in a relatively short time, mostly for the welding of both similar and dissimilar aluminium and other soft alloys. The need for rigid, lightweight structures in the engineering industries is driving further development of FSW of dissimilar materials. New exciting opportunities have opened for the welding of steels and other hard alloys although their commercial applications will require development of cost-effective and durable tools and probably process modifications.

The difficulties in FSW are easy to recognise but developing a vision for the future to overcome these is much more difficult. Perhaps the research may be directed in areas where conventional welding techniques have disadvantages, particularly focusing on the issues which are "show stoppers". In situations where fusion welding cannot be used because of incompatibilities of properties of the components, it may be necessary to bear the cost of expensive tools. The search for robust tools continues because there is a huge opportunity to develop FSW tools for the welding of hard alloys in order to successfully compete against the well-established fusion welding processes. FSW is a product of research and development in a mature field that is critical to support our standard of living. It is sometimes argued that progress in these fields is likely to be incremental because the theories and experiments have been tested over a long time. This diminished expectation of the outcome of organised research in welding and other mature areas is without any merit and is harmful to society. Only in a mature field such as welding that has so much impact on our standard of living, can an innovation provide so much societal benefit in such a short time.

3.9 Terminology

Fusion welding	Welding where the parts are melted locally and the solidification of the molten pool forms the welded joint.
Grain size	A normal metal is polycrystalline, i.e., it contains many crystals, often referred to as *grains*. Grain size is the mean size of the grains.
Heat treatable Al-alloys	Additional strengthening obtained by heating an appropriate alloy followed by rapid cooling to a lower temperature to induce the formation of fine crystals. This is known as artificial ageing.
Momentum	Product of mass and velocity, characterised by both magnitude and direction.
Non-heat treatable alloys	Aluminium alloys whose strength results from alloying elements and cold deformation.
Plastic deformation	Deformation of alloys that are irreversible.
Plasticised alloys	Hot solid alloys that flows under pressure. Precipitates: A compound that separates from the bulk alloy because of cooling.
Strength	Ability of alloys to resist deformation.
Strengthening	Enhancing strength of alloys by techniques such as adding alloying elements, having precipitates in the structure or cold working the alloy.
Solidus temperature	Temperature at which first liquid forms when an alloy is heated from low temperature.
Tracer material	Inert material added deliberately to an alloy to monitor its flow pattern.
Torque	Tendency of a force to rotate an object.
Traverse force	Force along the direction of motion of the tool to overcome the resistance of the material through which the tool traverses.
Toughness	Ability of an alloy to absorb energy and plastically deform it without fracturing.
Worm holes	A cavity in the friction stir welded joint below the surface.

References

1. W.M. Thomas, E.D. Nicholas, J.C. Needham, M.G. Murch, P. Temple-Smith, C.J. Dawes, Friction stir butt welding: International Patent Application No. PCT/GB92/02203, 1991
2. NASA completes welding on massive fuel tank for first flight of SLS rocket: https://www.nasa.gov/centers/marshall/about/star/star161005.html. Accessed June 2020
3. A. Arora, T. DebRoy, H.K. D.H. Bhadeshia, Back-of-the-envelope calculations in friction stir welding—velocities, peak temperature, torque, and hardness. Acta Mater. **59**, 2020–2028 (2011)
4. H.K.D.H. Bhadeshia, T. DebRoy, Critical assessment: friction stir welding of steels. Sci. Technol. Weld. Join. **14**, 193–196 (2009)
5. O. Frigaard, O. Grong, O.T. Midling, A process model for friction stir welding of age hardening aluminium alloys. Metall. Mater. Trans. A **32**, 1189–1200 (2001)

Chapter 4
Picture to Parts, One Thin Metal Layer at a Time

4.1 Introduction

Imagine a component that is so convoluted and with such intricate detail that it would be impossible to manufacture, or impossible to manufacture in one go. Figure 4.1 illustrates the problem; machining this out of a solid would simply not be possible. Precision casting by a process in which the mould can be burnt-off or removed mechanically might do the trick but would involve many manufacturing steps and considerable skill. It could be made in parts and then joined together to create the overall shape followed by intricate machining to get the dimensions and surface topology as required. All this amounts to a nightmare, great expense and a lot to time to achieve the finished product.

Now conduct a thought experiment in which the object is cut into thin, parallel slices without any loss of material. The patterns in each slice are captured on a computer, which then is used to drive a metal-printer which recreates each slice, overlays them in sequence, resulting ultimately in the three-dimensional shape in more or less its final form. This can be repeated as many times as is necessary. This algorithm can itself be digitally created using engineering design tools. It is important to realise in this process, that the method used for printing is less important than achieving the shape and properties in the final structure.

A simple chess piece, the castle, can be made much more elegant and desirable if it contained structure within, rather than just a solid lump illustrating the concept of a castle. Figure 4.2 shows how the 3D printing process can create internal features with re-entrant angles that would be *impossible to make* by any other process with the precision required. The chess piece suddenly becomes an item that enthrals. The skill in making this object is first in creating the concept, then in expressing it digitally and in choosing the right material for the job, in this case a ceramic which communicates rigidity and quality which a plastic in general would not. Furthermore, the chess piece is unlikely to be subjected to stresses and strains, so a metal is not necessary either.

© Springer Nature Switzerland AG 2021
T. DebRoy, H. K. D. H. Bhadeshia, *Innovations in Everyday Engineering Materials*,
https://doi.org/10.1007/978-3-030-57612-7_4

Fig. 4.1 An abstract metal
object that is made depositing
thin layers on top of each
other in a sequence that
reproduces the desired shape

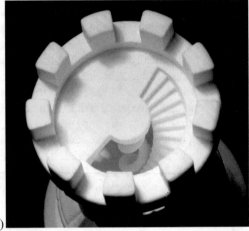

(a)   (b)

Fig. 4.2 The ceramic chess piece that is a castle, with a spiral staircase leading to the battlement,
accompanied by a spiral hand-rail at the centre

An additive manufacturing facility can in principle switch at a moments notice
to the production of a different part once a digital design is available. This can have
life-changing consequences. The COVID-19 pandemic led to a massive shortage of
components for ventilators and equipment for the protection of medical staff. AM
facilities in many parts of the world switched to the production of these vital goods,
sometimes with innovative design features. In fact, the rate limiting factor was not
the ability to manufacture but to get approval for the use of equipment in medical
scenarios.

Fig. 4.3 Bruce Wilmore, commander of the international space station showing a ratchet wrench made using a 3D printer [1]

The International Space Station is resupplied at regular though long intervals compared with the normal shopping practices on Earth. When a ratcheting socket wrench was required, a digital file was sent electronically by NASA to a desktop AM machine located in the orbiting space station. The machine then built a wrench including the movable parts, all in one piece, layer by layer in 104 layers from polymers. On completion, an astronaut simply retrieved the wrench and used it with good effect, Fig. 4.3. This was the first time an object designed on Earth was manufactured in space.

Metallic foams are used in sound damping, to provide rigidity in structures and have an advantage over polymeric foams in that they resist fire. They can form substrates for gaseous reactions. Their density can be less than that of water. Some applications require specific pore-geometries which present ideal opportunities in additive manufacturing. Figure 4.4 shows an open-pore metallic foam created in this manner. A metallic hollow-sphere is also illustrated—such spheres can have enormous specific strength (strength divided by density) so they can be used in lightweight construction with the advantage that additive manufacturing permits the design of such spheres as opposed to the acceptance of geometries limited by conventional manufacturing processes.

Figure 4.5 shows the distribution of revenues from consumer products, automotive, health care, aerospace, marine and other industries. The viability of these products in the context of AM depends on the ability to manufacture more easily than using conventional methods. Nevertheless, if the component is sufficiently sophisticated, then it becomes possible to manufacture it more rapidly than conventional methods given the avoidance of excessive machining, assembly and inspection.

Fig. 4.4 Additively manufactures metallic hollow-sphere and open-pore foam

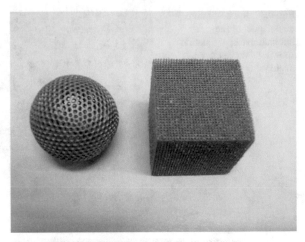

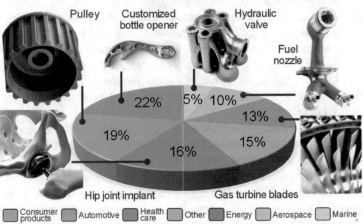

Fig. 4.5 Examples of metal printing in diverse industries. Each colour indicates an industry sector, i.e., consumer products, automobile, health care and other industries. The pictures represent a part in an industrial sector. For example, GE's fuel nozzle is a part used in aerospace industry. The revenue from all printed parts in the aerospace industry is 13% of the total revenue of all metal printed parts from all industrial sectors. Currently, the revenue generated by metal printing is growing but small compared with the total manufacturing industry [2]

4.2 Advantages and Disadvantages

Some of the aspects of the additive manufacturing process have been described above, but it is useful to present the big picture before dwelling into detail, Table 4.1.

Table 4.1 Some of the advantages and shortcomings of the additive manufacturing process

Advantages	Disadvantages
Complex shapes that are difficult to produce by conventional manufacturing processes can be printed	Productivity is limited because components are created in a stepwise manner. Resolution is limited by this step size (defined by powder size, layer thickness), both in the raster pattern and the thickness of the two-dimensional slice
Economical use of material given that final finishing of the product is all that is required	Surface roughness can present challenges—for example, the cooling channels in jet engine turbine blades can be less than 1 mm in diameter
Relatively small capital investment in manufacturing equipment	High productivity requires banks of printing machines
Ability to rapidly switch to production of different components	Residual stresses and distortion can compromise metallic components
The print-on-demand model can be implemented	Limited availability and limited variety of metallic powders or wires at affordable cost. In contrast, the alloys available for conventional manufacture have many orders of magnitude greater variety and lower cost
Customised parts can be produced without new tooling or equipment	
Functionally graded components possible by using different "inks" during the creation of a single component	Structural integrity of critical metallic components is difficult to achieve.
Rapid prototyping	Metallic components usually have less than 100% density

4.3 Metal Printing Processes

The heat sources required for metal deposition include continuous-wave carbon dioxide lasers, solid-state Nd:YAG lasers, Yb fibre lasers, electron beams and electric arc. Alloy powders or wires are commonly used as feedstock. A moving heat source melts an alloy layer by layer which subsequently solidifies. The trajectory of the heat source is determined from the algorithm for the part to be manufactured.

In the powder bed fusion process, shown schematically in Fig. 4.6a, the bed is lowered by a small distance after each layer is deposited, a roller spreads a thin layer of powder over the part and its surrounding area before another layer of the alloy is deposited with the help of a laser or electron beam [3]. In the directed-energy deposition system shown in Fig. 4.6b, the powder is supplied by a powder feeder co-axial with a laser beam. Both the laser beam and the powder feeder move relative to the part. The powder particles are heated during their flight and after they impinge on the part. After each layer of metal is deposited, the substrate is lowered slightly so that the distance between the heat source and the deposition surface does not change. A laser beam, an electron beam, or an electric arc can be used as a heat

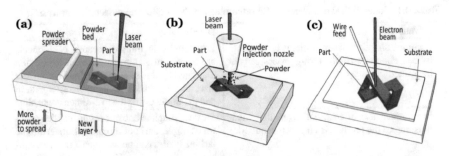

Fig. 4.6 Illustration of common additive manufacturing processes for metallic materials: the powder bed fusion process on the left, the directed-energy deposition system in the middle and the electron beam AM process on the right [3]

Table 4.2 Comparison of additive manufacturing processes for metallic components. The direct deposition processed samples may require hot-isostatic pressing to reduce porosity and grinding or machining depending on the finish required. The components produced using electric arc welding usually require machining. Components made using powder bed fusion rarely require hot-isostatic pressing

Process	Direct energy deposition			Powder bed fusion
Feedstock	Powder	Wire		Powder
Heat source	Laser	Electron beam	Electric arc	Laser or electron beam
Power/W	100–3000	500–2000	1000–3000	50–1000
Speed/mm s^{-1}	5–20	1–10	5–15	10–1000
Feed rate/g s^{-1}	0.1–1.0	0.1–2.0	0.2–2.8	–
Size/m × m × m	2 × 1.5 × 0.75		5 × 3 × 1	0.5 × 0.28 × 0.32
Production time	High	Medium	Low	High
Dimensional accuracy/mm	0.5–1	1–1.5	2–5	0.04–0.2
Surface roughness/μm	7–20	8–15	Needs machining	4–10

source for the directed-energy deposition process. Either a stream of powder or a wire feed can be used to build components.

Table 4.2 shows the main features of the common metal printing processes which can help select the optimum method for the intended component. They all take considerably longer times to build parts than casting or injection moulding. A larger powder particle-size or wire diameter limits the resolution possible, making it more difficult to achieve fine features. The powder bed process uses finer powders to deposit intricate features; greater scanning speeds are also possible compared with the direct energy deposition method. Consequently, the powder bed fusion method is associated with less heat input per unit length of the deposit, the layers are much thinner, and they cool much faster. Smaller laser spot diameters and finer powders allow better control of geometric features in the parts. Slower deposition rates produce improved surface quality using thinner layers at the expense of productivity.

When large parts are made they tend to retain heat for a longer duration and the cooling rates are relatively slow. Often, items larger than 30 × 30 × 30 cm are produced in near net-shape by melting of a wire followed by machining. High deposition rates are achieved by simultaneously using two wire electrodes. Wire based techniques that use welding power sources are gaining popularity, partly because of the ready availability of wires that would normally be used in welding technologies.

4.4 Uniqueness of Printed Parts

The printing of metals is attractive because it can produce components that cannot be easily and economically produced by conventional manufacturing. Jet engine fuel nozzles are now made routinely in a custom-designed factory [4]. In the past, each fuel nozzle was an assembly of about twenty individual parts. They are now partially manufactured by laser melting of alloy powders, layer by layer, in 20 μm thin layers (one-fifth of the thickness of a human hair). The fuel nozzle, shown in Fig. 4.7a, reduces the number of assembly steps required and it is claimed to be five times as durable as the conventional nozzles.

Metal printing offers additional opportunities to create components with site specific chemical composition and properties. The nuclear industry often uses joints between steel and nickel alloys, the latter being more resistant to elevated temperature exposure. An abrupt change from one to the other causes huge changes in the vicinity of the joints, in particular, the partitioning of carbon on to the steel side where it locally embrittles the structure. This abrupt change can be mitigated by designing a joint where the chemical composition varies gently from the steel to nickel alloy. This can easily be achieved using directed energy deposition to create a compositionally graded joint as illustrated in Fig. 4.7b. All that is needed is to feed different proportions of the component powders during deposition.

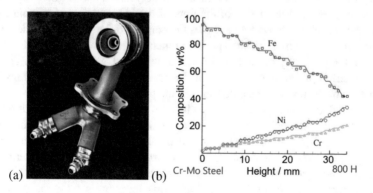

(a) (b)

Fig. 4.7 (**a**) The tip of the GE jet engine nozzle made by the AM process [4], made by laser powder bed fusion, in one step. (**b**) Compositional profile of a functionally graded joint between a nickel alloy 800H and a Cr–Mn steel [5]. The points are measured values of concentrations in layers directly deposited using a laser as the energy source. Image adapted from [5] with the permission of Elsevier

Metal printing now routinely makes customised products such as patient-specific implants and legacy products where the supply chain no longer exists.

4.5 Underlying Principles

The physics associated with additive manufacturing process affects the microstructure, properties, and the ability to service the manufactured components. Metal printing involves heating and melting of the feedstock, followed by solidification of the liquid and then cooling in the solid state. In the direct energy processes, the feedstock receives heat even before reaching the build surface and may absorb heat during flight before impinging on the deposition surface, which also receives heat directly from the source. The powder or wire therefore melts quickly and the molten pool is propagated along the predetermined track.

The highest temperature on the melt pool surface is attained directly below the heat source and then decreases with distance from this location. Surface tension is a function of temperature so its variation with position creates the so-called Marangoni stress on the surface of the molten pool, which makes the liquid move from regions of low to high surface tension.

The three-dimensional flow of liquid metal in the pool is important because it affects both the dissipation of heat and the mixing of the feedstock with the molten metal from the pre-existing layers. However, it is difficult to determine the motion experimentally since liquid metals are opaque and the pool is small and moves rapidly. A recourse is to simulate metal flow by numerically solving the equations of conservation of mass, momentum and energy with initial and boundary conditions [5, 6]. This approach of simulating liquid metal velocities in a computer rather than by direct experimental measurement is widely adapted in engineering practice. Most of our current knowledge of the flow of liquid metal in AM originated from numerical modelling. Figure 4.8a shows the computed flow pattern inside a molten pool during powder bed fusion. The liquid metal moves away from under the heat source to the periphery of the liquid pool, turns around and recirculates. The speed and orientation of circulation determine the extent of convective heat transfer and the mixing of the hot and the cold fluids in the pool.

The circulation pattern obviously influences the temperature distribution in the liquid alloy, its heating and cooling rates, solidification pattern, and the evolution of various solid phases that make up the microstructure. The solidification morphology, grain structure and the phases that form define the microstructure and the mechanical properties of the printed component.

An important consequence of building parts layer by layer is the temperature excursion that each location of the part experiences. Unlike most other materials processing operation, in AM each location of the part experiences multiple temperature peaks. For example, Fig. 4.8b shows the computed thermal cycles at various monitoring locations inside a part. Temperatures at the mid-height and mid-length of several layers are shown as a function of time. The first temperature peak

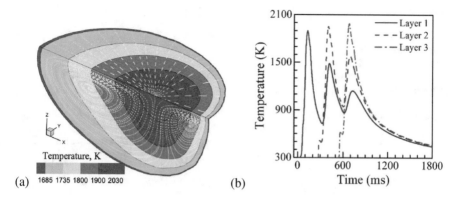

Fig. 4.8 (**a**) Circulation pattern in the fusion zone. Each streamline shows the path for the flow of liquid metal [3]. (**b**) Calculated thermal cycles in a laser assisted direct energy deposition of stainless steel at a laser power of 210 W and 12.7 mm s^{-1} speed [6]. The monitoring locations are at the mid-height and mid-length of each layer

corresponds to a position of the laser beam just above the monitoring location. The subsequent peaks occur during the deposition of the upper layers. So, at each location, the microstructure and the grain structure of the alloy that forms after the first thermal cycle are often changed by the subsequent thermal cycles depending on the specific temperatures and times. These thermal cycles affect the evolution of microstructure and the eventual mechanical properties of the part.

The substrates used in AM are effective heat sinks. As a result, the peak temperatures attained in the lower layers close to the substrate are somewhat lower than those in the upper layers [6]. In the upper layers the distance from the heat sink increases and the peak temperature rises because of the reduced heat loss. So, the thermal cycles are inherently spatially dependent. An important consequence of this result is the microstructural asymmetry of the part. Since the structure affects properties, the properties may also be inherently different at different locations. Each location in the component experiences multiple thermal cycles and phase transformations, grain growth, residual stresses, distortion that affect the mechanical properties of the component.

4.6 Evolution of Structure and Properties

The structure and properties of alloys depend on their thermal histories that are affected by many variables such as the scanning speed, power, power density, scanning pattern, part geometry and the thermo-physical properties of the alloy. All these variables affect heat transfer within the part, which control temperature profiles and cooling rates. Given the many causative variables and their wide range of values, it is no surprise that the cooling rates reported in the literature for the

Fig. 4.9 Reported cooling
rates for the printing of
stainless steel for a wide
variety of processing
conditions [2]. (**a**) Powder
bed fusion using laser. (**b**)
Directed energy deposition
using laser. (**c**) Direct energy
deposition using electric arc

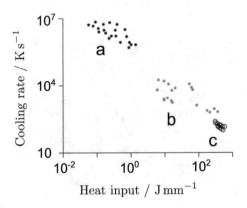

printing of a stainless steel in Fig. 4.9 show a wide range as a function of processing variables [2]. The heat input, i.e., the energy deposited per unit length, has a seminal influence of the cooling rate. The wide range at the same time provides an opportunity to customise the cooling rate for specific properties. The cooling rate may not of course be uniform throughout the component, which will lead to corresponding variations in properties that must be taken into account in the design process.

The morphology of the solidification structure is affected by the temperature field, which in turn depends on the scanning pattern used during deposition. Consider the two scanning patterns, one where the heat source travels along the same direction, i.e., always scanning from left to right and another where the direction alternates between left to right followed by right to left, in the context of the deposition of a nickel base alloy, Fig. 4.10 [7]. Solidification occurs by the epitaxial growth from the substrate of columnar dendrites, with growth direction influenced by that in which heat flows, which in turn depends on the scan pattern. The orientation of the dendrites is identical in all layers when the scan direction is maintained constant. In contrast, alternating the scan direction causes corresponding changes in the dendrite orientations between adjacent layers [8]. The orientations of crystals are important in determining properties because if they are all similarly oriented, they may not, for example, have the same fracture properties as when they are differently oriented (i.e., differently *textures*). Therefore, the deposition sequence matters as illustrated in Fig. 4.10. There are other properties, such as hardness, that are influenced by the chemical composition of the deposit. If the composition of a particular class of nickel alloys is expressed empirically in terms of a single parameter ϕ:

$$\phi = w_{Ni} + 0.65 w_{Cr} + 0.98 w_{Mo} + 1.05 w_{Mn} + 0.35 w_{Si} + 12.6 w_C$$

$$- 6.36 w_{Al} + 3.80 w_B + 0.01 w_{Co} + 0.26 w_{Fe} + 7.06 w_{Hf} + 1.20 w_{Nb}$$

$$+ 4.95 w_{Ta} + 5.78 w_{Ti} + 2.88 w_W$$

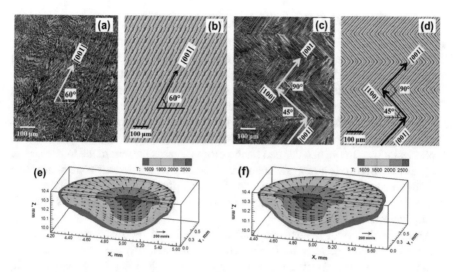

Fig. 4.10 Influence of heat source travel direction on the growth pattern of primary dendrites in nickel alloy [7]. (**a**) Dendrite growth direction when laser beam traverses left to right for all layers. (**b**) Computed dominant heat flow direction corresponding to (**a, e**). (**c**) Alternating dendrite growth direction as the scanning orientation reverses in successive layers, with corresponding information in (**d, f**)

Fig. 4.11 The dependence of Vickers hardness on chemical composition of nickel alloys. Various processing conditions were used. For details and the limitations of the analysis, see [7]

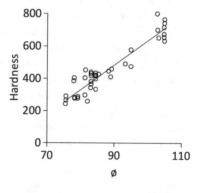

where w_i is the weight percent of solute i, then ϕ is found to correlate strongly with the hardness of the deposit, Fig. 4.11. An exciting opportunity will arise when sufficient data on additive manufacturing are openly available so that they can be subjected to machine learning techniques to reveal quantitative patterns that can be used to make further advances. Such work has been developed using neural networks, for the powder bed fusion based on electron beams as the energy sources, to estimate the strength of the deposits [9].

4.7   Defects and Other Challenges

Porosity, lack of fusion, solidification cracking, residual stresses and distortion are issues that need to be improved upon for the qualification of printed parts. When alloys contain volatile elements, some might selectively evaporate leading to uncontrolled changes in the intended chemical composition.

Several mechanisms are responsible for the porosity formation. The feedstock may contain dissolved gases that are evolved during solidification to form small, spherical pores. In the powder bed fusion process, gases present in the inter-particle spaces may become entrapped. The very high-temperature vapour zone beneath an electron or laser beam may collapse due to instability of the power density or local changes in powder packing, leading to porosity when solidification occurs after the beam has left the locality.

Adjacent layers of deposit may not fuse together if the fused region does not penetrate the surface to a sufficient depth. Insufficient overlap between adjacent tracks in the scan or between layers may leave unmelted regions in between. So the appropriate choice of deposition conditions is vital to ensure a dense solid, and the conditions may differ with the types of material deposited. Figure 4.12a shows macroscopic defects that arise during the powder bed process due to a lack of fusion during deposition. In Fig. 4.12b, successive layers separate because the thermal contraction stresses that are not homogeneously distributed exceed the strength of the interface between the layers.

Solidification cracking occurs when the tensile stress due to volume shrinkage associated with the liquid→solid transformation, is such that during cooling, the shrinkage stresses exceed the strength of the solidifying region the elevated temperatures involved. The composition of the alloy, geometry of the deposited bead and scanning speed can all affect this type of cracking. Irregular cracks, up to a few millimetres in size can be generated. This type of cracking depends on the chemical composition of the alloy, particularly when it contains impurities such that the temperature range over which a mixture of solid and liquid is extended.

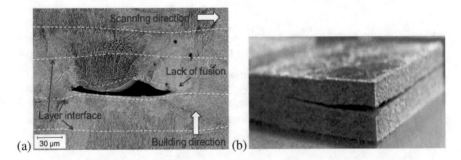

Fig. 4.12 (a) Lack of fusion defect in a nickel alloy during powder bed fusion [8], reproduced with permission of Elsevier. (b) Delamination in stainless steel deposits during powder bed fusion using a laser heat source [10], image courtesy of the University of Texas

Table 4.3 Important variables that affect distortion of components additively manufactured. Other factors such as the gap between successive layers, periodic stress relief and the volume change of transformation also affect the development of residual stress

Variable	Mechanism	Remedy
Heat input, i.e., the power of the heat source divided by the scan speed	Larger heat input results in a greater pool of liquid and greater temperatures. Non-uniform cooling results in shrinkage and distortion	Reduce heat input. Smaller liquid pool results in reduced shrinkage and distortion
Temperature coefficient of volume expansion	Larger volume shrinkage during cooling makes an alloy more susceptible to distortion	Thermal management to reduce peak temperature and cooling rate
Rigidity of the alloy and constraint	Rigidity describes the ability of material to resist deformation, the product of stiffness and geometric factors	Higher stiffness of a material and/or a thicker plate can resist distortion

Residual stresses are those that exist in a body even at equilibrium. They evolve during spatially non-uniform heating and cooling of the metal, thermal expansion and contraction, phase transformations and uneven distribution of plastic strains. Mitigation strategies include control of the substrate preheat temperature, shorter deposition length or scanning in smaller bits, more rapid scanning and thinner layers. A large preheat temperature can also be helpful in preventing solidification cracking. Table 4.3 lists a few methods to reduce the effects of residual stresses.

Most engineering alloys typically contain one or more volatile alloying elements. Manganese and chromium in stainless steels, magnesium and zinc in aluminium alloys and aluminium in titanium alloys are examples. Their selective evaporation changes the chemical composition, which may or may not be an issue. A reduction in peak temperature by selecting an appropriate power distribution pattern, higher heat source power that results in smaller surface-to-volume ratio of the molten pool, and faster scanning speed will minimise the problem.

4.8 Concluding Remarks

A distinguishing feature of additive manufacturing is its ability to create internal features such as cooling channels within an otherwise solid part. If deposition leads to distortion and surface roughness, then these internal features may not be as intended in the original design. A smaller liquid pool size with lower heat input can often help, but this must inevitably reduce productivity and increase cost.

Additive manufacturing is obviously a layer-by-layer process, so curved surfaces are approximated in a stepwise profile. A smaller step size is needed for a smoother macroscopic curvature, but it may ultimately be necessary to smooth the surface mechanically. Tiny metal drops and unmelted or partially melted powder particles ejected from or near the fusion zone (i.e., spatter) by high-speed metal vapours and gases, often land on the build surfaces and contribute to roughness. Large unmelted powder particles or "balls" are often found at the edge of the molten pool. High heat input may reduce the severity of the problem by melting large particles, but small powder particles also result in a smoother surface. Powders as fine as 20 μm have been used to ensure a better surface finish.

The scientific challenges include the interrelation between processing, microstructure, properties and performance, microstructure control, minimisation of defects, and poor solidification and grain structure. A better understanding can eliminate some of the trial-and-error used in fixing AM parameters. Similarly, rapid qualification of parts, overcoming geometric limitations, scaling-up, printing sequence, and the health and safety concerns of handling fine metal particles, are examples of technological problems. Commercial challenges include cost competitiveness, availability of feedstock and a need for standards.

To facilitate more rapid printing of large parts, multiple heat sources and wire feed mechanisms are explored, together with hybrid methods that use a combination of manufacturing technologies.

4.9 Terminology

Alloying elements	Elements in an alloy added to enhance its properties.
Columnar dendrite	Columns of tree like solid structures that form during solidification of liquid alloys.
Conduction	Mechanism of heat transfer in a stationary solid or liquid due to temperature difference.
Convection	Movement of liquids or gases. If hot gases and liquids are in a motion, they can carry significant amounts of heat with their motion.
Electron beam	A stream of energetic electrons capable of heating and melting alloys in a focused area.
Epitaxial growth	Atomic arrangements of a new growth layer conforming to the structure of the existing layer.
Laser beam	A device that can emit an intense beam of light through stimulated emission of radiation. A focused laser beam can melt and vaporise alloys.
Marangoni convection	Flow of liquids from low to high surface tension regions.
Microstructure	The magnified pattern of a surface observed using a microscope.
Solidification morphology	Shape of the solids that form from the liquid alloys.
Surface tension	A measure of how closely molecules on the surface stick to each other.

References

1. 3D printed wrench: 2015: http://www.collectspace.com/news/news-122914a-3D-printer-space-station-ratchet.html
2. T. DebRoy, T. Mukherjee, J. Milewski, J. Elmer, B. Ribic, J. Blecher, W. Zhang, Scientific, technological and economic issues in metal printing and their solutions. Nat. Mater. **18**, 1026–1032 (2019)
3. T. DebRoy, H. Wei, J. Zuback, T. Mukherjee, J. Elmer, J. Milewski, A. Beese, A. Wilson-Heid, A. De, and W. Zhang, Additive manufacturing of metallic components–process, structure, and properties. Prog. Mater. Sci. **92**, 112–224 (2018)
4. 2020: https://www.ge.com/additive/stories/new-manufacturing-milestone-30000-additive-fuel-nozzles
5. J.S. Zuback, T.A. Palmer, T. DebRoy, Additive manufacturing of functionally graded transition joints between ferritic and austenitic alloys. J. Alloys Compd. **770**, 995–1003 (2019)
6. V. Manvatkar, A. De, T. DebRoy, Melt pool geometry, peak temperature, and solidification parameters during laser additive manufacturing. Mater. Sci. Technol. **31**, 924–930 (2015)
7. H.L. Wei, J. Mazumder, T. DebRoy, Evolution of solidification texture during additive manufacturing. Sci. Rep. **5**, 16446 (2015)
8. M. Xia, D. Gu, G. Yu, D. Dai, H. Chen, Q. Shi, Porosity evolution and its thermodynamic mechanism of randomly packed powder-bed during selective laser melting of Inconel 718 alloy. Int. J. Mach. Tools Manuf. **116**, 96–106 (2017)
9. P.C. Collins, C.V. Haden, I. Ghamarian, B.J. Hayes, T. Ales, G. Penso, V. Dixit, G. Harlow, Progress toward an integration of process–structure–property–performance models for "three-dimensional (3-D) printing" of titanium alloys. J. Metals **66**, 1299–1309 (2014)
10. K. Kempen, L. Thijs, B. Vrancken, S. Buls, J.V. Humbeeck, and J. Kruth, Producing crack-free, high density M2 HSS parts by selective laser melting: pre-heating the baseplate, in *24th International Solid Freeform Fabrication Symposium*, Arizona (Laboratory for Freeform Fabrication, Austin, 2013) pp. 131–139

Chapter 5
Welding: The Digital Experience

Metal pieces can be joined together without riveting or using mechanical means. One of the methods by which this can be achieved is known as *fusion welding*. This method is in fact so popular that there are vast and critically important engineering structures assembled by locally melting the edges and using additional molten metal to fill the gap. Figure 5.1 shows the world's tallest bridge, the Millau Viaduct in France, in which the huge deck-sections were welded together on site, using about 150 tonnes of filler metal.

Like many technologies, the early days of fusion welding involved human creativity, perseverance and a considerable amount of experimentation. That metal could be melted using an electric current was patented in England during 1865 [1]; in 1886 Elihu Thompson obtained patents for the fusion welding process and demonstrated that metals could be welded [2]. At the time, this latter patent was said to be the *foundation of the new art* [2]. So why was it referred to as an art? Why do the most modern reviews of the subject still refer to the state-of-the-art rather than the state-of-the-science, which strangely, is the phrase describing reviews in the humanities subjects such as management and psychology?

A cursory examination of the complexity of the process reveals why. Figure 5.2 shows a snapshot of just a few of the phenomena that must be captured in order to make welding into a science. The electrode has a carefully calibrated chemical composition which on transfer into its final state will not only be compatible with the metal being joined, but also accounts for losses due to the volatility of some elements. The arc that is struck between the electrode and metal reaches up to 20,000 °C, in which case a plasma is created with the structure of the arc varying from the hottest region at the centre to the relatively cooler parts at its periphery. The transfer of metal must be protected so an inert gas shield is implemented, which itself can have consequences on the ultimate shape of the weld pool. Droplets can detach from the electrode in a variety of ways, for example, spray transfer, dip transfer, etc.

© Springer Nature Switzerland AG 2021
T. DebRoy, H. K. D. H. Bhadeshia, *Innovations in Everyday Engineering Materials*,
https://doi.org/10.1007/978-3-030-57612-7_5

Fig. 5.1 Millau Viaduct, the tallest bridge in the world, built using concrete and steel in harmony. Photograph courtesy of Yaan de Carlan. The structure is an epitome of elegance which would not have been possible without welding. The final phase of the construction, i.e., the joining together of the leading edges of the north and south decks, involved welding at a height of 270 m above the Tam valley. The deck itself has a complex internal structure that ensures rigidity and fatigue resistance

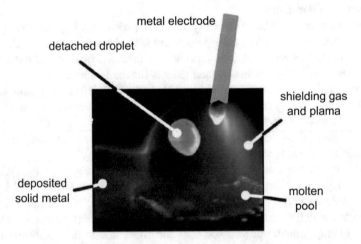

Fig. 5.2 A snapshot from a high-speed video recording of electric arc welding using a wire electrode. The arc is struck between the electrode and the steel. This melts the consumable electrode with droplets detaching and falling into the weld pool below—some of the steel being joined also melts and contributes to the weld pool while protected by a shield of inert gas. Image courtesy of Stuart Guest, Julien Chapuis and Patricio Mendez

The pool of molten metal underneath is not static, but subject to complex flows within and to deformation by forces such as surface tension, arc forces and buoyancy forces. The heat source is of course the electric arc, but the way in which the energy is dissipated determines the efficiency of the process. Once solidified, the metal cools, perhaps permitting some gases to escape from solution or in the case of hydrogen, loss by diffusion into the surrounding steel where it can do damage. The cooling process itself leads to structural changes within both the pool and the solid steel in its vicinity. Stresses develop due to the heterogeneous implementation of the heat in a localised region of a controlled assembly. Every aspect of metallurgy, physics, chemistry and engineering is involved in the creation of a weld.

So it is not surprising that the adjective *art* persists given that there is as yet no complete science to deal with all the issues described, and many more, in a single integrated model. Nevertheless, there has been quite dramatic progress.

5.1 Moving Heat Source

Although Fig. 5.2 illustrates the electric arc welding process, which is by far the most common technique used in practice, there are extensive applications of other heat sources such as laser and electron beams. So the power density, speed and penetration abilities can be selected at will. There is another process, resistance spot welding where two pieces of metal can be joined by pressing copper electrodes from either side, with resistance heating at the gap between the sheets; this process is different in the sense that the heat source is stationary. However, in most applications, the heat source or work piece is translated in order to create a lengthy weld, Fig. 5.3.

It is important to know the thermal cycle experienced by the work piece due to the moving heat source, because this cycle amounts to a heat treatment that alters the structure of the adjacent metal that is not melted, i.e., the *heat-affected zone* of the weld. Rosenthal derived an analytical equation to describe the temperature field around the heat source [3]:

$$T - T_0 = \frac{q}{2\pi \kappa v} \exp\{-\lambda v \psi\} \exp\{-\lambda v R\}/R \qquad (5.1)$$

where T is the temperature, T_0 the temperature prior to welding, κ the thermal conductivity, λ the thermal diffusivity, v the welding speed, $\psi = x - vt$ where x is the coordinate along the welding direction, t is the time, $R = \sqrt{\psi^2 + y^2 + z^2}$ where y is the coordinate normal to the welding direction in the plane of the plate being welded and z is normal to the plate. This equation embodies the arc power q and the heat input per unit length q/v. The equation was very influential in capturing the essence of the quantitative treatment of the thermal consequences of welding but has the following approximations which Rosenthal himself recognised:

- The heat is provided as a *point source*, whereas in practice it is a distributed source.
- There is no melting treated. There is no filler metal.

Fig. 5.3 The electrode
moves along the welding
direction which is towards the
right, leaving behind the
solidified weld metal. There
are many variations on this
process [4, 5]. Image courtesy
of Kamellia Dalaei, ESAB
AB

- The far-field temperature of the work piece is unchanged.
- Suppose that the melt pool shape is calculated by assuming that it is defined by
  the surface where the temperature exceeds the melting temperature. It would be
  calculated as having a semi-circular section in the yz plane.

Given these stimulating limitations and the importance of the pool shape in actually
achieving a sound joint with sufficient penetration, there are now much more
sophisticated treatments of the problem, albeit requiring a computer to solve the
fundamental equations repeatedly.

5.2   The Real Weld Pool

The weld pool is not a quiescent liquid, rather there are circulation patterns due to a
combination of phenomena. The shape can therefore vary from a fairly flat pool to
one shaped like a keyhole.

   An important factor is due to Carlos Marangoni, who demonstrated how liquid
flow is driven by a non-uniform surface tension with flow directed towards regions
of high tension. In the case of a weld pool, the surface temperature is not
homogeneous and since surface tension is a function of temperature, circulation
patterns are set up within the depth of the pool. Gravity causes the colder, heavier,
liquid near the edge of the weld pool to sink and the lighter liquid metal in the
middle of the weld pool to rise. In arc welding an electromagnetic force is generated
from the interaction between the current path in the weld pool and the magnetic
field it generates. So, the motion of the liquid metal results from a combination of
Marangoni, gravitational and electromagnetic forces.

   Of these three forces, that due to gravity is by far the weakest; during arc
welding, the electromagnetic force is comparable to the Marangoni force only at
high currents. In most cases, the Marangoni force dominates the flow of metal within
the weld pool. The rolling streams of weld metal carry heat from underneath the
heat source to all other locations within the weld pool. Its circulation determines
the melting pattern of the various regions of the work piece, the shape and size
of the weld pool, and the structure and properties of the welded joint. These flow

patterns can only be revealed by computation of the vector fields, but their ultimate consequences are vivid in the shape of the weld pool and in some cases, the distribution of solute in the solidified weld.

5.2.1   Many Billion Equations

Both the temperatures and velocities at all locations in the weld pool affect the shape and size of the weld pool, the mixing of the filler metal, cooling rates at different locations, the extent to which alloying elements vaporise and hence are lost to the pool, thus affecting the weld metal composition, and the structure and properties of the joint. The local temperatures and velocities can be calculated by solving equations representing the conservation of mass, momentum and energy [6]. The task is complicated so the object is partitioned into some 250,000 connected cells for numerical representation. For each cell, an algebraic equation relates the local values of a variable with its values at the neighbouring cells [7]. Typically the variables include three components of velocities, enthalpy or temperature, and pressure which are solved repeatedly, until closure is obtained. For these five variables, a total of $5 \times 250,000$ or 1.25 million equations have to be solved for each iteration. And many thousands of iterations are needed before correct solutions for the variables at all cells are obtained. Today, a billion such linear algebraic equations can be solved in about 2 min using inexpensive computers.

Figure 5.4 shows for an instant of time, the computed temperature and velocity fields during an arc weld. Since the heat source is moving, the temperature changes rapidly in the cold work piece ahead of the moving weld pool. Regions of different temperatures are shown by specific colour bands. These regions are compressed in front of the weld pool which is on the left of the picture. Behind the weld pool where the material has already been heated, the metal cools slowly in air and the temperatures change more gradually.

Fig. 5.4 Computed flow of weld metal during arc welding. The colours represent temperatures in Kelvin and the dotted lines the flow of liquid. The two loops shown near the surface are caused by the Marangoni effect, whereas those below the surface result from electromagnetic force [6]

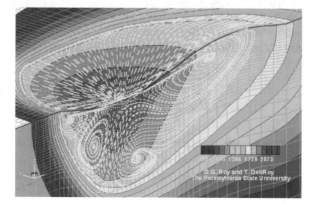

Fig. 5.5 The heat-affected zone highlighted by the oxide interference colours on a welded stainless steel tube. Photograph courtesy of Drs Yanhui Zhang and Simon Condie of TWI, Cambridge

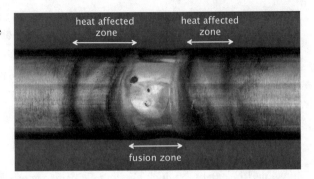

It evident that the computed surface of the pool is not flat because liquid metal moves away from the low surface tension region under the heat source to other regions where the tension is higher. It is depressed below the arc which exerts pressure on the liquid surface forming a small hump behind the arc. The velocities range from a few tens of centimetres per second to about a metre per second, and the liquid metal carries a significant amount of heat from under the heat source to other locations within the weld pool.

The method also allows the temperature fields to be computed once the weld solidifies and in the heat-affected zone. Figure 5.5 shows the typical extent of the material in the vicinity of the weld that is influenced by the heat input during the welding. The most profoundly affected regions are approximately of the same width as the fusion zone.

The computed thermal information can then be fed into mathematical models founded on thermodynamic and kinetic theory, that enable the structure and properties to be estimated [8]. In the case of steels, it is possible for a huge variety of crystals to grow from the red-hot state, some of which are beneficial and others are not. Figure 5.6a shows some of the key crystals of importance in low-alloy steel weld deposits; of these, acicular ferrite is desirable because it improves the ability of the weld metal to absorb energy during fracture, a key feature in ensuring the safety of engineering structures. Figure 5.6b illustrates how this kind of a microstructure can be calculated as a function of the welding conditions that determine the thermal profile of the weld metal as it cools, and the chemical composition of the metal. Similar calculations can also be implemented for the heat-affected zone [9].

5.3   Mystery Resolved

There was a puzzle at one time, that welds made on the same grade of steel showed quite different weld pool shapes for identical welding conditions [10], Fig. 5.7. The essential difference was minute variations in the sulphur concentration, frequently regarded as an impurity.

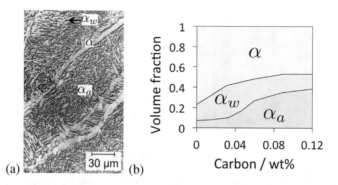

(a)   (b)

Fig. 5.6 (**a**) Scanning electron micrograph of the primary microstructure of a steel weld (courtesy of Rees). The terms α, α_w and α_a refer to allotriomorphic ferrite, Widmanstätten ferrite and acicular ferrite, respectively. (**b**) Variations in microstructure and mechanical properties as a function of carbon concentration in Fe-1Mn-C wt% steel weld deposit using manual metal arc welding $(1\,\mathrm{kJ\,mm^{-1}})$

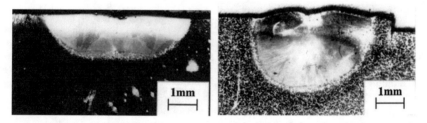

Fig. 5.7 Weld cross-sections of steel plates containing 20 ppm sulphur (left) and 150 ppm sulphur (right) spot welded at a laser power of 5200 W for 5 s [11]. The chemical compositions otherwise were not different

A number of scientists suggested a role of sulphur or selenium in affecting the weld pool shape [7, 12, 13]. These elements affect the way in which the surface tension of liquid steel depends on temperature, Fig. 5.8. At small concentrations of sulphur, the surface tension decreases relative to pure iron, and shows a different trend, i.e., increasing with temperature. At very high temperatures close to the boiling point, the surface tension then decreases as the temperature increases. Sulphur and other solutes such as oxygen, nitrogen, selenium and tellurium have a tendency to migrate to the surface of the liquid steel. They all affect the surface tension in a manner similar to sulphur and are known as surface active elements [14].

Directly under the heat source, the liquid metal has the highest temperature and lowest surface tension when the steel contains practically no sulphur. Since liquids flow from low to high surface tension regions, hot liquid steel moves sideways from the middle of the weld pool to its edge and melts metal there. It then turns downward as shown in Fig. 5.9a. As a result, the weld pool becomes wide and shallow.

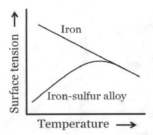

Fig. 5.8 Variation of surface tension with temperature [14]. For pure liquid iron, the surface tension decreases with temperature. When a small amount of sulphur is added to liquid iron, the surface tension drops, and then increases with temperatures until the temperature reaches close to the boiling point

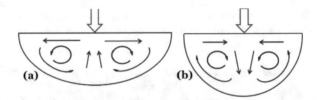

Fig. 5.9 (**a**) Pure iron flows sideways from the middle making the weld pool wide and shallow. (**b**) When a small amount of sulphur is added, the alloy goes downward in the middle of the weld pool resulting in a deep weld pool

Small additions of sulphur change the flow pattern completely. Hot liquid under the heat source now has a higher surface tension than those in the cooler regions, Fig. 5.8. So, on the surface of the weld pool the weld metal rushes to the middle and then moves downward to the bottom of the weld pool. The downward flow of the hot metal in the middle of the weld pool works like a thermal drill and a deep weld pool forms as shown in Fig. 5.9b.

This hypothesis provided a plausible explanation. Tiny particles of alumina were added to the liquid pool and their trajectories at the surface were monitored using special cameras, which revealed the expected change in flow pattern due to sulphur [15].

When the sulphur content is 150 ppm, the circulation pattern is opposite to what was observed for the 20 ppm sulphur steel weld shown in Fig. 5.10b. The surface velocities are fairly large, higher than $20 \, \text{cm} \, \text{s}^{-1}$. So the heat is carried mostly by convection. Hot weld metal flows downward under the heat source, the base metal melts near the root and a deep weld pool forms.

But sulphur does not always change the weld pool geometry. No perceptible difference in the cross-sections of low power laser welds containing 20 and 150 ppm of sulphur in steel is shown in Fig. 5.11. The numerical simulation of heat transfer and fluid flow reveals why.

The computed results reveal lower peak temperatures and lower velocities in the weld pool for these small welds. Convection did not carry much heat since the

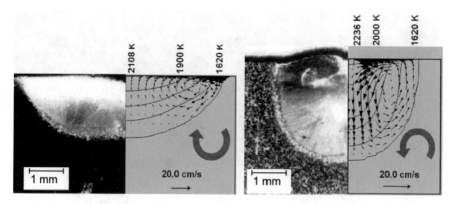

Fig. 5.10 Experimentally determined and theoretically calculated weld pool geometries in a 15 mm thick high speed steel plate spot laser welded for 5 s. The welds had 20 and 150 ppm sulphur on the left and right sides, respectively [11]

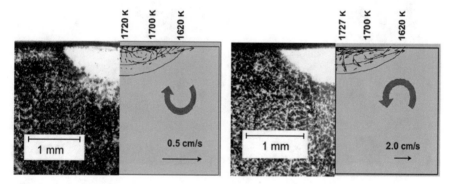

Fig. 5.11 Comparison of the computed and experimental weld pool geometries at a laser power of 1900 W for steels containing (**a**) 20 ppm and (**b**) 150 ppm sulphur [11]

velocities involved were small, making conduction the main mechanism of heat transfer. Therefore, sulphur had little effect on the weld profile.

5.4 Joint Orientation

The shape and size of the fusion zone affect the mechanical properties of the joints. Analytical methods such as that in Eq. 5.1 cannot predict such effects, but numerical methods come into their fore when dealing with the multitude of phenomena involved. Figure 5.12 shows sections of the simulated weld normal to the welding direction, in one case for a symmetrical V-shaped joint and in the other case when the V-shape is tilted as illustrated. In both cases, the considerable depression of the free surface due to the arc force is pronounced when the heat source is directly

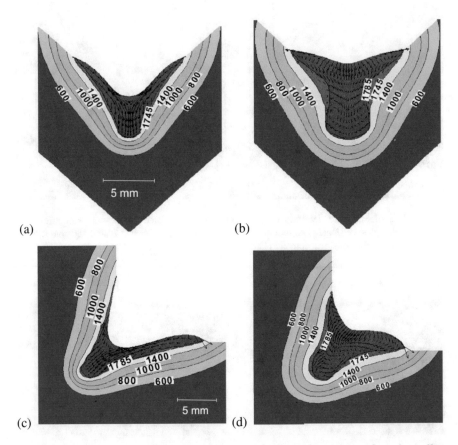

Fig. 5.12 Numerical heat transfer and fluid flow models capture how joint configurations affect the fusion-zone shape and size during gas metal arc welding [16]. (**a, c**) represent a location directly below the heat source, whereas (**b, d**) are for 5 mm behind the heat source. The welding speed in all cases was $100\,\mathrm{cm\,s^{-1}}$, the temperatures are in Kelvin and the fusion-zone boundary is assumed at 1745 K

above the fused region. The liquid metal flows downwards in the middle of the pool, driven by the electromagnetic force, causing a finger-like penetration. When the heat source has moved away, the situation 5 mm behind the arc is illustrated in Fig. 5.12b, d. The arc pressure is smaller, causing a correspondingly smaller weld pool depression. When the V-joint is tilted during welding to an L-configuration, the free surface shows asymmetry due to gravitational force both directly under the arc and behind the arc.

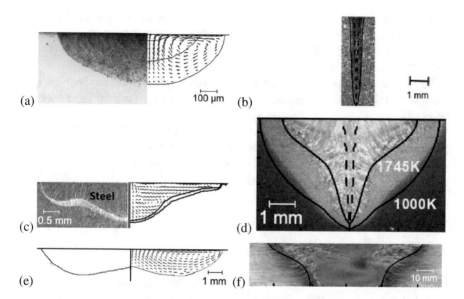

Fig. 5.13 A variety of weld shapes, both experimental and corresponding calculated shapes. The details are given in [17–20]. (**a**) Aluminium alloy, arc weld, with actual shape on the left and that calculated on the right. (**b**) Electron beam weld on stainless steel—the line represents the calculated shape and the dashed line the keyhole shape. (**c**) As (**a**) but for the laser welding of sodium nitrate. (**d**) Laser and arc hybrid weld in stainless steel. (**e**) As (**a**) but laser weld on steel. (**f**) Friction stir weld on aluminium alloy

5.5 Weld Geometry

Figure 5.13 shows the variety of weld pool shapes as a function of welding process and the material being welded. It is evident that in all cases, there is an accurate modelling of the shape. The shape can vary from approximately hemispherical to a deep weld when a laser of electron beam is used as the heat source. In the latter cases, a *keyhole* may form during welding, which is essentially a vapour cavity that is surrounded by the liquid pool. A combined laser and arc source naturally results in a hybrid shape as well (Fig. 5.13d).

5.6 Summary

Mechanistic models of welding clearly can empower engineers to embed the most advanced of scientific understanding into practice. They have been shown capable of capturing the relationships between the heat source, welding speed, temperature and velocity fields, cooling rates, and solidification growth rate, all based on interdisciplinary scientific principles. Many of the important phenomena associated with welding cannot in fact be measured, for example, due to the opacity of the weld pool. They can also define experiments when the calculations do not measure up to data. Furthermore, the method can reduce the resources required to reach the desired outcome that is specified by manufacturing constraints.

The models are more often than not, of generic value; many of them are now being adapted to guide 3D printing where components are built up layer by layer (Chap. 4).

There is in all this, an educational role. Students are inspired by calculations that make predictions that can subsequently be verified using critical experiments—after all, this is the best practice scenario in all of the natural sciences.

5.7   Terminology

Alloying elements	Elements in an alloy added to enhance its properties.
Conduction	Mechanism of heat transfer in a stationary solid or liquid due to temperature difference.
Convection	Movement of liquids or gases. If hot gases and liquids are in a motion, they can carry significant amounts of heat with their motion.
Electron beam	A stream of energetic electrons capable of heating and melting alloys in a focused area.
Laser beam	A device that can emit an intense beam of light through stimulated emission of radiation. A focused laser beam can melt and vaporise alloys.
Marangoni convection	Flow of liquids from low to high surface tension regions.
Microstructure	The magnified pattern of a surface observed using a microscope.
Surface tension	A measure of how closely molecules on the surface stick to each other.

References

1. P.W. Fassler, Twenty five years of electrical welding. Amer. Welding Soc. J. **10**, 29–35 (1935)
2. A.M. Dunlap, A.H. Munch, An investigation of electric welding. Thesis for bachelor of science in electrical engineering, University of Illinois, Illinois, 1911
3. D. Rosenthal, Mathematical theory of heat distribution during welding and cutting. Weld. J. Res. Suppl. **20**, 220s–234s (1941)
4. J.F. Lancaster, *Metallurgy of Welding*, 4th edn. (Allen and Unwin, London, 1986)
5. S. Kou, *Welding Metallurgy*, 2nd edn. (Wiley, New Jersey, 2003)
6. T. DebRoy, Role of interfacial phenomena in numerical analysis of weldability, in *Mathematical Modelling of Weld Phenomena II*, vol. 1005 (The Institute of Materials, London, 1995), pp. 3–21.

7. S. Kou, D.K. Sun, Fluid flow and weld penetration in stationary arc welds. Metall. Trans. A **16**, 203–213 (1985)

8. H.K.D.H. Bhadeshia, L.-E. Svensson, B. Gretoft, Model for the development of microstructure in low alloy steel (Fe-Mn-Si-C) weld deposits. Acta Metall. **33**, 1271–1283 (1985)

9. J.C. Ion, K.E. Easterling, M.F. Ashby, Diagrams of microstructure and hardness for HAZ's in welds. Acta Metall. **32**, 1949–1962 (1984)

10. T.J. Lienert, P. Burgardt, K.L. Harada, R.T. Forsyth, T. DebRoy, Weld bead center line shift during laser welding of austenitic stainless steels with different sulfur content. Scr. Mater. **71**, 37–40 (2014)

11. W. Paischeneder, T. DebRoy, K. Mundra, R. Ebner, Role of sulfur and processing variables on the temporal evolution of weld pool geometry during multikilowatt laser beam welding of steels. Weld. J. **75**, 71s–80s (1966)

12. C.R. Heiple, J.R. Roper, Effect of selenium on GTAW fusion zone geometry. Weld. J. **60**, 143s–145s (1981)

13. W.S. Bennett, C.S. Mills, CTA weldability studies on high manganese stainless steel. Weld. J. **53**, 548s–553s (1974)

14. P. Sahoo, T. DebRoy, M.J. McNallan, Surface tension of binary metal surface active solute. Systems under conditions relevant to welding metallurgy. Metall. Trans. B **19**, 483–491 (1988)

15. C.R. Heiple, J.R. Roper, Mechanism for minor element effect on GTA fusion zone geometry. Weld. J. **61**, 97s–102s (1982)

16. A. Kumar, T. DebRoy, Heat transfer and fluid flow during gas-metal-arc fillet welding for various joint configurations and welding positions. Metall. Mater. Trans. A **38**, 506–519 (2007)

17. A. Arora, G.G. Roy, T. DebRoy, Unusual wavy weld pool boundary. Scripta Metall. **60**, 68–71 (2009)

18. B. Ribic, R. Rai, T. DebRoy, Numerical simulation of heat transfer and fluid flow in GTA/laser hybrid welding. Sci. Technol. Weld. Join. **13**, 683–693 (2008)

19. R. Rai, P. Burgardt, J.O. Milewski, T.J. Lienert, T. DebRoy, Heat transfer and fluid flow during electron beam welding of 21Cr-6Ni-9Mn steel and Ti-6Al-4V alloy. J. Phys. D Appl. Phys. **42**, 025503 (2009)

20. A. Arora, R. Nandan, A.P. Reynolds, T. DebRoy, Torque, power requirement and stir zone geometry in friction stir welding through modeling and experiments. Scr. Mater. **60**, 13–16 (2009)

Chapter 6
Inventions that Enabled the Silicon Age

6.1 Introduction

Silicon transistors, since their discovery in the 1950s, have penetrated the lives of practically all human beings on Earth. Smart phones, tablets, computers, kitchen appliances, supercomputers and other devices too numerous to mention are all founded on silicon-based electronics. The computers have enabled machine control, machine learning, self-driving cars and space-craft, the Internet and have blurred the differences between science fiction and reality. Not apparent from these spectacular achievements are the particular innovations without which the silicon age would not have dawned.

Jöns Jacob Berzelius, a Swedish chemist, discovered amorphous silicon, a solid with a haphazard arrangement of atoms, in 1824, Fig. 6.1. Some 30 years later, the French chemist Henri Sainte-Claire Deville made the first crystalline silicon. These are the only two forms of silicon that exist. Metallurgical grade silicon with a minimum of 98 wt% purity has been produced in bulk since the 1850s. Alloys of silicon are used for deoxidation of steel, since silicon has a greater affinity for oxygen than iron, so by reacting with the dissolved oxygen in the liquid, it refines the steel. Aluminium-silicon alloys are used for casting automotive engine blocks. Polymers that contain silicon are used in non-stick cooking utensils, electrical and household sealants and numerous other applications. However, the metallurgical silicon is not pure enough to make a transistor or associated electronic devices. Impurities greatly affect the mobility of charge carriers in electronic materials such as silicon and its performance in devices.

Silicon for semiconductor devices needs to contain less than one impurity atom in ten million silicon atoms, commonly known as nine-nine purity, i.e., 99.9999999% pure. Small amounts of particular elements such as boron or phosphorous can then be added deliberately to change its electrical properties. It then can conduct electricity, or become an insulator depending on conditions. Silicon semiconductors are used for the construction of transistors that can amplify electrical signals or act

© Springer Nature Switzerland AG 2021
T. DebRoy, H. K. D. H. Bhadeshia, *Innovations in Everyday Engineering Materials*,
https://doi.org/10.1007/978-3-030-57612-7_6

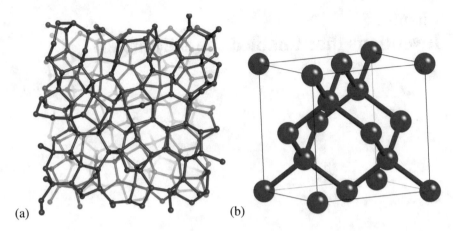

(a)   (b)

Fig. 6.1 Distribution of atoms in: (**a**) amorphous silicon showing tetrahedrally bonded atoms in a random arrangement. Picture courtesy of Matthew Cliffe [1]. (**b**) Crystalline silicon where the atomic arrangement has long-range periodicity so that all the atoms in the crystal can be generated by stacking the cubic cell in all directions to fill space. Notice that the arrangement of atoms is identical to that of diamond (Fig. 2.1) but the nature of the bonding between silicon atoms is very different

as switches that are the building blocks for almost all electronic devices we use on a daily basis.

6.2   The Source of Silicon

Silicon is widely available in the earth's crust but it is the quartzite that is mined for industrial production. Quartzite forms by the transformation of quartz-rich sandstones under the influence of high temperature and pressure. It is a durable, hard rock and has a chemical formula of SiO_2. Figure 6.2a shows a piece of quartzite containing faceted crystals of silica. Its colour varies, ranging from grey to pink depending on the small concentrations of impurities such as iron oxide, alumina and calcium oxide. China, Russia, the USA and Norway are the biggest producers of silicon (Fig. 6.2b) and contributed about 82% of the global production of 6.7 million metric tonnes of silicon in 2018.

6.3   Pathway to Electronic Grade Silicon

Figure 6.3 shows the sequence involved in the two commonly used routes in the production of silicon. Metallurgical grade silicon is converted to either trichlorosilane ($SiCl_3H$) or silane (SiH_4) and these gases are subsequently purified by fractional

Fig. 6.2 (**a**) A piece of quartzite rock that shows silica crystals of varying colour because of the differences in the concentration of impurities. Courtesy of the Mineral Museum, College of Earth and Mineral Sciences, Penn State University. (**b**) Amount of silicon produced in 2018 in various countries [2], expressed in 1000 metric tonnes of silicon. The total production figures used includes silicon metal as well as a common alloy of iron and silicon known as ferro-silicon

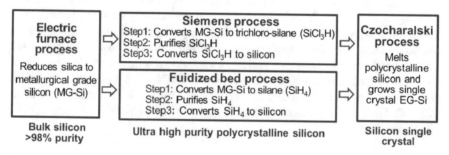

Fig. 6.3 Alternative paths for making very high-purity single crystal silicon from the metallurgical grade silicon containing about 98 wt% Si and small amounts of C, Fe, Ca, Ti, Cr, O, B and P. EG-Si refers to high-purity electronic grade (EG) silicon used in the manufacture of electronic devices, whereas MG-Si is the less pure metallurgical grade (MG)

distillation. High-purity polycrystalline silicon (poly silicon) is then made from silicon-bearing gases, traditionally by the Siemens process but more recently using fluidised bed technology. The final step of converting polycrystalline silicon into to single crystal form by the Czochralski (CZ) process is common to both routes.

6.4   Production of Metallurgical Grade Silicon

Figure 6.4 illustrates schematically an electric furnace used in the production of metallurgical grade silicon by the reduction of quartzite ore with carbon. Typical electric furnaces are about 6 m in diameter and 3 m deep, lined with carbon bricks. They typically produce about 12 tonnes of liquid silicon per day.

The charge consists of quartzite and various forms of carbon such as coke, coal and wood chips. The carbonaceous charge serves as a reducing agent and enables

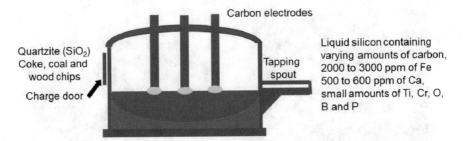

Fig. 6.4 A schematic diagram of an electric furnace for the manufacture of metallurgical grade silicon. Electric arc and resistive heating generate heat in the furnace. Temperatures vary spatially but peak values reach about 2100 K. Typical productivity is about 12 tons of silicon per day with a recovery of about 70% of the silicon in the charge

the passage of electrical current through three carbon electrodes. Typically, about 1100 kg of 98% purity liquid silicon is tapped periodically from the furnace. The size of the ore and carbonaceous materials, their purity and electrical resistance are important factors for the smooth operation of the process. The temperature is not uniform within the furnace so a variety of chemical reactions occur depending on the local temperature which can be as high as 2100 °C. The compounds that form include SiO gas and SiC which is a refractory material with a melting point of 2830 °C. The following chemical reactions occur in the production of metallurgical grade silicon:

$$SiO_2 + C = SiO + CO$$

$$SiO + 2C = SiC + CO$$

$$SiC + SiO = 2Si + CO$$

Since carbon is present in the electrodes, furnace crucible and coke, the silicon produced contains about 500 parts per million of carbon by weight. Depending on the nature and concentration of impurities present in the furnace charge, the molten metal also contains 2000–3000 ppm iron, 1500–4000 ppm of aluminium, 500–600 ppm of calcium and smaller quantities of titanium, chromium, oxygen, boron and phosphorous. The power consumption is about 14 kW-h per kg of silicon for the operation of a furnace that produces 12 tonnes of metallurgical grade silicon per day.

6.5 Manufacture of Pure Polycrystalline Silicon

6.5.1 Siemens Process

Developed in the 1960s, the Siemens process converts the impure silicon into volatile $SiCl_3H$ by reacting it with HCl gas in a fluidised bed reactor at about 300 °C:

$$Si + 3HCl = \underbrace{SiCl_3H}_{trichlorosilane} + H_2 \qquad (6.1)$$

The reaction is favoured at low temperatures and low partial pressure of hydrogen. The chloride is distilled to remove impurities. The exhaust from the fluidised bed reactor is filtered to separate unreacted silicon particles, and any HCl in the exhaust is absorbed in a sodium hydroxide solution and the gas is then liquified below -30 °C in a condenser. The trichlorosilane is then separated from the liquid by distillation (Fig. 6.5).

The separation of pure silicon from trichlorosilane is achieved by heating to ≈ 1150 °C after mixing it with hydrogen in a chamber containing silicon rods which are electrical-resistance heated, essentially a reversal of the reaction outlined in Eq. 6.1. Silicon deposits on these rods. This process takes more than 200 h and is energy intensive. After completion, the polycrystalline rods of silicon are crushed and melted to make cylindrical single crystals in a crystal growing facility.

Silicon with a maximum of one part per billion of impurity by weight is produced by the Siemens process. It is a simple and dependable process and its operating parameters have been optimised over many decades; most electronic grade silicon is now made in this way. However, there is significant cost when the plant is shut down periodically extract the silicon rods after they have grown to a certain diameter. The

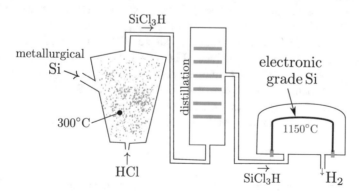

Fig. 6.5 Schematic illustration of the Siemens process. In the final stage, polycrystalline silicon is deposited on to the silicon rods that are electrically heated. A typical reactor may contain several dozen silicon rods. The steel bell jar is cooled. The cost of production is between $9 and $14 per kg. Adapted from Alba Ramos [3]

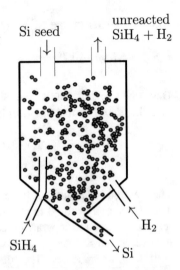

Fig. 6.6 A schematic diagram of the fluidised bed reactor to produce high-purity silicon

process also requires special care in the handling of the corrosive hydrochloric acid gas. Most important, it is energy intensive, in part because of the wasted heat in cooling the reactor walls. There is no clear pathway to improve energy efficiency of the Siemens process. The fluidised bed process is now emerging as a commercially viable alternative.

6.5.2   Fluidised Bed Process

This involves the thermal decomposition of silane into silicon and hydrogen. The process currently contributes less than 10% of the global production of high-purity polysilicon. As a continuous process, it is an attractive alternative because it uses less than 20% of the energy needed for the Siemens process.

There are various ways of producing silane SiH_4, but its production from trichlorosilane is the most economical. Figure 6.6 shows a simplified diagram to produce silicon in a hot walled fluidised bed reactor where granular silicon seed particles are suspended in a mixture of silane and hydrogen. Silicon from the silane decomposition (Eq. 6.2) deposits on the seed particles in the range 650–850 °C which is much lower than the operating temperature for the Siemens process.

$$SiH_4 = Si + 2H_2 \qquad\qquad (6.2)$$

The silicon seed particles continue to grow through deposition from the decomposing silane. The larger particles of polysilicon are removed periodically from the bottom of the reactor and new seeds charged from the top. Some amount of silicon does not deposit on the seed particles and forms dust in the reactor. Minimising dust formation remains an important goal to prevent its build up inside the furnace. The exhaust from the furnace contains hydrogen, some silicon particles and the unreacted silane.

There are several advantages of the fluidised bed process compared with the Siemens method. The operating temperature is lower, a greater energy efficiency, chloride-free, less corrosive feed gas, all of which contribute to a lower cost of producing polysilicon. The purification of silane is much easier than that of trichlorosilane because of the significant difference in the relative volatilities of silane and the common impurities. Energy loss due to cooling of the bell jar (Fig. 6.5) in the Siemens process is avoided. However, fluidised beds are more complex to operate than the Siemens bell jar with static silicon rods. Leakage of silane and hydrogen could be hazardous. The purity of silicon produced by the Siemens process, higher than nine nines purity, which is purer than that produced in the fluidized bed reactors (six to nine nines). However, the Siemens process has been optimised during many years of operation but it is possible that the fluidised bed process will make inroads in the coming decades [4].

6.6 Silicon Single Crystal by Czochralski Process

The polycrystalline silicon rods and granular silicon pieces are converted to crystalline silicon by the Czochralski process. High-purity silicon is placed in a pure silica crucible and melted under an inert gas environment. The melt is heated to a temperature just over the melting point of silicon. As shown in Fig. 6.7, a seed crystal of silicon is introduced in the melt and slowly pulled up while gently rotating the crystal. The crucible is also rotated in the opposite direction to reduce crystal imperfections. At the start of the process, the seed crystal is pulled relatively fast and the cross-section of the crystal is small. After a short time, the speed of pulling is reduced to increase the diameter to that desired.

The quality of the crystals depends on the temperature of the liquid, solidification velocity and the rotational speeds of both the crystal and crucible containing liquid silicon. These variables are used to reduce the temperature gradient in the crystal. The spatial variation of temperature within the crystal may result in local stresses and the generation of dislocations. Since the solubilities of impurities are higher in the liquid than in solid, the impurities are rejected by the solidifying crystals into the liquid. As a result, the liquid gets progressively enriched with impurities during crystal growth. Therefore, either more polycrystalline silicon is added in the crucible or the entire liquid is not used for growth and the process is terminated with still some amount of liquid intentionally left in the silica container.

The single crystals made by the Czochralski process have diameters of 450 mm and weigh up to 300 kg. They are then sliced into 0.6–0.8 mm thick wafers using an annular circular saw. A circular hole in the saw serves as the cutting edge on which small diamond particles are embedded. Alternatively, many parallel wires between 100 and 200 μm in diameter coated with small diamond particles are used to

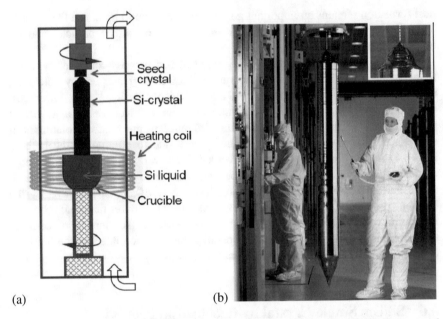

(a)   (b)

Fig. 6.7 (**a**) Silicon crystal growth arrangement by the Czochralski process. (**b**) A silicon crystal grown by the CZ process. Image © 2000 KAY CHERNUSH — ALL RIGHTS RESERVED

Fig. 6.8 Single crystal silicon wafer, 800 mm in diameter, placed in a reactor for the manufacture of light-emitting diodes. The surface is reflective so the features inside are simply reflections. Photography courtesy of Saptarsi Ghosh and Rachel Oliver, The Cambridge Centre for Gallium Nitride

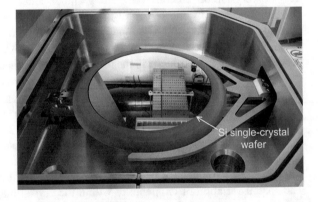

make wafers (Fig. 6.8) from crystals grown by the Czochralski process. Wafers are processed further to make microprocessors and numerous other electronic devices. Silicon single crystals of extreme purity, largely free of crystal imperfections are crucial for the production of most electronic devices.

6.7 Concluding Remarks

We have seen that the manufacture of silicon for electronic devices requires certain levels of purity and crystal perfection. The Siemens process consistently yields extreme purity silicon that can be applied in the majority of silicon-based devices. The fluidised bed is an emerging technology which currently produces silicon that has a very high purity, but less than that from the Siemens process; nevertheless, the silicon from the fluidised bed can be used in the manufacture of devices such as the photoelectric cells used for solar power.

6.8 Terminology

Condenser	A device that cools gases so they turn into a liquid.
Coke	Produced by heating coal in the absence of air to improve its strength. Coke is often used as a reducing agent in furnaces.
Deoxidation	Removal of undesirable oxygen from a molten metal. Carried out by adding materials that have a high affinity for oxygen.
Distillation	Purification of liquid by heating, vaporising and cooling or condensing. Different liquids can be separated from a mixture by boiling off the components at different temperatures.
Fluidised bed	Reactor where solid particles are suspended in a flowing gas to facilitate rapid heat transfer and chemical reaction between the gas and the solid.
Microprocessor	Integrated assembly of transistors and other electronic components that processes inputs according to programmed instructions.
Outer shell	Contains the electrons on an atom that determine its ability to interact with other atoms.
Polycrystalline	A material made up of many crystals in different orientations.
Polysilicon	High-purity polycrystalline silicon.
Quartz	Crystalline mineral SiO_2.
Semiconductor	Neither a good conductor nor a good insulator.Its electrical conductivity can be changed by doping or with the application of an electrical field.
Silane	Chemical compound of silicon and hydrogen used in the production of high-purity silicon.
Single crystal	Has long range periodicity with no discontinuities such as grain boundaries.
Transistor	An electronic device that works as an amplifier of electrical signals or as an electrical switch. Often made of silicon or germanium with small amounts of dopants to manipulate electrical properties.
Unpaired electron	Electron in energy level that is not balanced by another opposite spin.

References

1. F. Wooten, K. Winer, D. Weaire, Computer generation of structural models of Si and Ge. Phys. Rev. Lett. **54**, 1392–1395 (1985)
2. U. S. Department of the Interior, *Mineral Commodity Summaries*. (Available from Government Publishing Office, Washington, 2019)
3. A. Ramos, Understanding and improving the chemical vapor deposition process for solar grade silicon production. Ph.D. Thesis, Universitat Politècnica de Catalunya, 2016
4. Anonymous, Making the chips that run the world. Smithson. Mag. **30**(10), 36 (2000)

Chapter 7
Transition to Sustainable Steelmaking

7.1 Introduction

Products made from steel are all-pervasive. Agricultural machinery, housing, distribution systems for clean water, roads, transportation, generation of energy and an endless list of all the things rely on steel, that add to the quality of life. However, the current technology of making of steel is energy intensive and devastatingly polluting, discharging about 1.83 tonnes of carbon dioxide into the environment per tonne of steel produced [1]. In 2018 the global production of steel was about 1.7 billion tonnes [2] associated with the discharge of 3.1 billion tonnes of carbon dioxide into the environment. Steel production therefore contributed to 8.4% of the total of 37.1 billion tonnes [3] of carbon dioxide pumped into the atmosphere during 2018. Given the global population growth and the rising standard of living, the demand for steels is likely to rise in the foreseeable future and the increased steel production will intensify the emissions.

The consequences of CO_2 emissions are well established; urgent action is required over the next 7 years to avoid the irreversible rise in temperature above the $2\,^{\circ}C$ higher than that during the start of industrialisation, after which the rise in sea levels, and more chaotic weather systems will play havoc with humanity.

After many decades of research and development into alternative steel production technologies, it is blindingly obvious that investment into new green technologies which are able to cope with the huge demand, and legislation to reduce emissions using sustainable steelmaking may be the only remaining pathway. A monetary cost associated with CO_2 production may need to be levied on both the producers and consumers of steel.

Possible replacement of the current steelmaking technology that has been perfected over a century by green innovations is not straightforward. Capital investment in promising steel-technologies is difficult because of the low profit margins of the steel industry; we currently pay less for steel than for bottled water, soda or coffee in a weight-for-weight comparison. The long-time horizon for a return on investment

© Springer Nature Switzerland AG 2021
T. DebRoy, H. K. D. H. Bhadeshia, *Innovations in Everyday Engineering Materials*,
https://doi.org/10.1007/978-3-030-57612-7_7

and the inherent high risk in any large scale, complex technological venture also add to the difficulties. Any new technology must be suitable for mass production. To appreciate the scale, consider the time needed to produce the iron for the four gigantic steel cables that support the 1.45 km long George Washington Bridge that connects New Jersey with New York. They weigh about 22,300 tonnes and all the steel for these cables can be produced in about 2 days from iron made a modern blast furnace. So, any new technology must be capable of supporting production on a very large scale in order to substitute for the current technology.

Here we explore the fundamental reason why the production of steel has such a large carbon footprint, the current status of less-polluting technologies, the implementations of policies needed for their deployment and the outlook for the future.

7.2 Why Does Steel Have Such a Large Carbon Footprint?

There are two main routes of steelmaking. The bulk of the steel is produced in basic oxygen furnaces from molten pig iron that is produced in blast furnaces. Secondly, it is also produced from recycled scrap steel in electric arc furnaces. Both routes contribute to the large carbon footprint.

Blast furnaces are charged with sintered pellets of haematite (Fe_2O_3) or magnetite (Fe_3O_4) ores, calcite or dolomite flux, coke and air (Fig. 7.1). Inside the blast furnace, the oxygen in the air reacts with carbon in coke to provide heat and form carbon monoxide (CO) which travels upward through the stack and reduces the iron oxides to generate iron and carbon dioxide. The carbon that originally was charged as coke leaves the furnaces as CO and CO_2, with some dissolved in the hot metal. Carbon dioxide also is generated by the decomposition of the limestone and/or dolomite that are necessary to form the slag, which dissolves many solutes that are undesirable in the molten pig iron. The CO_2 exits the furnace as a part of the top gas.

Pig iron (which has a high carbon concentration and relatively poor properties) is subsequently converted into steel in basic oxygen furnaces that are refractory-lined pear-shaped vessels. A supersonic jet of commercially pure oxygen is made to impinge on to the liquid metal bath to oxidise some of the carbon in the liquid iron. The oxidation of carbon, manganese, silicon and other alloying elements is highly exothermic resulting in intense heating. The CO produced by the oxidation of carbon burns at the top of the furnace to produce CO_2 in the furnace exhaust.

In an integrated process, the sintering plant where the charge is compacted for mechanical strength, coke ovens, blast furnaces and basic oxygen furnaces all contribute to the CO_2 tally. In addition, the combustion of fuels and recycled gases in sintering plants, coke ovens, stoves for heating air for blast furnaces and various furnaces for heating solid steel for rolling, forging and annealing contribute to the CO_2 emissions.

Fig. 7.1 (a) Iron ore sinter for blast furnace. (b) Coke for blast furnace

In electric arc furnaces, the heating and melting of recycled scrap steel begins with the application of large electrical currents through carbon electrodes; the heat required may be supplemented by gas burners. The concentrations of copper, chromium, nickel, molybdenum and carbon in the charge need to be carefully controlled. Carbon is added in the furnace if its concentration in the bath falls below the target composition. Provision is made for the oxidation of carbon and other alloying elements in the liquid bath by lancing with oxygen. During this refining, CO and CO_2 are generated from the oxidation of carbon in the melt and the carbon electrodes. Air is added to the hot waste gas to convert the CO to CO_2. Modern electric arc furnaces make about 150 tonnes of steel in about 90 min, but they consume huge amounts of electricity. Thus, the CO_2 emissions from the electric furnace steelmaking are generated both directly from the oxidation of carbon and burning of fuels in the furnace and indirectly from using large quantities of electricity which is usually sourced by burning fossil fuels.

Steels produced in both BOFs and EAFs are further refined to adjust their chemical compositions in ladle furnaces. In most cases they are then continuously cast and rolled or forged into finished products. Depending on whether the latest technologies have been implemented, the process may not be seamless through all the stages that lead to the final produce. Therefore, the continuously cast products would be stockpiled in some instances and then reheated for processing in rolling mills to produce sheets, bars, rails and other shapes. Both the reheating of the steels in gas fired furnaces and their rolling require energy and hence have carbon footprints.

In short, there are three main sources of CO_2 emissions from steel plants. First, sinter plants, coke ovens, blast furnaces, basic oxygen furnaces and electric arc furnaces, all of which are major producers of CO_2. Second, the combustion of natural gas and other fuels in facilities such as the hot blast stoves, reheating furnaces and gas fired heating of ladles contribute to the generation of CO_2. Finally, many units of steel plants such as the electric arc furnaces and rolling mills consume enormous quantities of electric power and contribute indirectly to the generation of CO_2, bearing in mind that most electricity is generated using fossil fuels.

7.3 Current Status of Green Technologies

The technological challenges of developing and implementing green technologies are significant since deep cuts in CO_2 emission during steel production will require radical concepts. One possibility is to find an alternative reductant. Natural gas is used already for the direct reduction of iron oxide ore, but this does not obviate the production of CO_2. Hydrogen for reducing the oxide also has an environmental cost if fossil-fuel generated electricity is used to electrolyse water. Green energy, if available, would be the ideal way forward, but is not available in the amounts needed in the steel industry. On the other hand, nuclear powered hydrogen generation might be feasible if there is a sufficient motivation, although the capital costs would be enormous, and the lead time to build the reactors would not solve the 2 °C problem in a timely manner.

Energy usage in the steel industry has decreased consistently over the last half a century and Fig. 7.2 shows that by 1960 standards, the consumption has decreased by about 60%. The current usage, 20 GJ tonne^{-1} of steel, is at an all-time low.

Many new concepts have been evaluated in laboratories in an effort to mitigate the CO_2 emissions associated with steel production. A few have progressed to pilot plant scale to test their practical feasibility. Here we discuss two promising technologies, top gas recycling blast furnace and Hlsarna ironmaking technology. The former retrofits existing blast furnace equipment and the latter does not require the manufacturing of coke and sinter. These are by no means the only promising technologies, but they probably are at a greater degree of technology readiness.

Fig. 7.2 Relative quantities of energy consumption per tonne of crude steel produced. Adapted using data from [1]

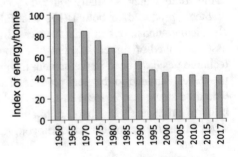

7.4 Top Gas Recycling Blast Furnaces

In the current blast furnace process, carbon is used both for the heating of the solid charge and the reduction of the oxide ore. Inside the furnace, oxygen in air reacts with coke to produce both heat and the reductant, carbon monoxide. Carbon dioxide is produced from the reduction of the iron ore, and any unreacted carbon monoxide and nitrogen leaves the furnace as exhaust gas. The current practice also heats N_2 in air which does not participate in furnace reactions and ends up in the exhaust gas.

In the proposed process which has been investigated since 1980 air is replaced by commercially pure oxygen to avoid heating large amounts of nitrogen. The exhaust gas is now a mixture of CO and CO_2. The CO_2 from the furnace affluent is separated from CO and appropriately stored. The separation of CO_2 is now easier, because its concentration in the gas mixture is large given the absence of the nitrogen in the exhaust gas. The remaining CO in the exhaust gas is recycled into the blast furnace. Figure 7.3 shows the main processes of the top gas recycling blast furnace. The changes from the traditional blast furnace process are both profound and ironic, because the furnace is still called the blast furnace without any "air blast". However, in the new process, the heating of nitrogen is eliminated, and the recycled CO reduces the consumption of coke in the furnace. Most important, the existing highly productive equipment, blast furnace, is retrofitted so that the investment required to implement the process is reduced.

The equipment used to separate CO_2 from the top gas operates on the principle of the capture of CO_2 in an absorbent, such as a zeolite. Subsequently CO_2 is separated from zeolite by reducing pressure in the absorption chamber. As shown in Fig. 7.3, after removing CO_2 from the top gas, a portion of CO is heated to 1173 K and introduced in the blast furnace through a new set of tuyers at the lower stack level above the original tuyers to achieve greater efficiency. The remaining CO is heated

Fig. 7.3 A schematic diagram of a top gas recycling blast furnace. Temperatures indicated are from [4]

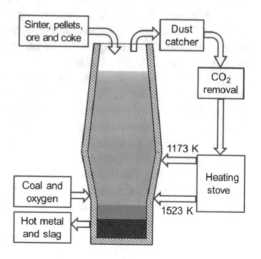

to 1523 K and injected into the furnace through the original tuyers into the hearth of the furnace. In addition, oxygen and coal, both at room temperature, are injected through the hearth-level tuyers. Because the recycled top gas contains CO, carbon input into the furnace is significantly reduced, typically from 470 to 350 kg per ton of hot metal.

The benefits of the process include the reduced consumption of coke; avoiding the discharge of CO_2 and using the existing blast furnace with retrofitted tuyers. Investment is needed for the equipment necessary to capture CO_2 and the installation of the new tuyers. There is of course the recurring cost of operating the absorption equipment and for the oxygen injected into the furnace. But there is no doubt about the environmental benefit. The concept has been tried in steel plants for extended periods at the pilot plant scale, but its deployment in larger production scale blast furnaces will require further development.

7.5   Hlsarna

Hlsarna is among the most promising of the new ironmaking technologies that has been studied at pilot plant scale in a sustained 6-month trial in 2017 in Tata steel's IJmuiden steelworks in the Netherlands. Its most attractive features include the elimination of several of the plants needed to make iron in the blast furnace. Coke ovens, pelletising and sintering plants are not needed to make iron in the Hlsarna process. About 20% reduction of energy usage, 20% lower carbon dioxide discharge and significant saving of manufacturing cost of liquid iron are attainable compared with the blast furnace process. It still needs to overcome the major technological challenges associated with its scale-up and subsequent establishment of its cost-competitiveness with the current blast furnace technology.

Hlsarna gets its name from the Celtic word for iron, Isarna, and Rio Tinto's smelting vessel, Hismelt. Since non-coking coal fines can be used, it eliminates the need for coke ovens. It uses ore fines, and therefore does not require iron ore pellets or sinters. The quality of the ore and all other raw materials is less stringent than those used in the blast furnace.

Figure 7.4 shows a schematic diagram of the Hlsarna reactor. Iron ore fine and oxygen are introduced in the upper part of the reactor where a high temperature cyclone facilitates rapid heating and melting of the iron ore. In this region, oxygen reacts with carbon monoxide to generate heat and produce carbon dioxide, which is withdrawn from the top of the furnace. The high concentration of the carbon dioxide in the exhaust gas facilitates carbon storage or use and further reduces the carbon footprint of Hlsarna. Powdered coal is introduced in the lower portion of the furnace where iron oxide is reduced to make liquid iron which settles at the bottom of the reactor. A layer of slag forms above the metal layer. Both the slag and the liquid metal are tapped periodically.

The current Hlsarna pilot plant can produce about 60 thousand tonnes of liquid iron per annum, which is about 60-fold lower than the typical production of a

Fig. 7.4 A schematic
diagram of Hlsarna reactor

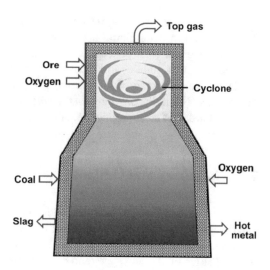

modern blast furnace. However, efforts are now underway to construct a larger plant with higher capacity. In summary, Hlsarna eliminates coke ovens, sintering and pelletising plants, has less stringent requirement on the quality of ore and coal, and generates carbon dioxide at high concentrations that facilitates its capture and reuse. These attributes and a sustained 6-month successful trial in a pilot plant scale makes Hlsarna an energy efficient, cleaner attractive technology. Its effectiveness as a business venture will depend on the success of its scale-up and the cost-competitiveness of the iron produced.

7.6   Overcoming Barriers Through Implementation of Policies

There is no well-tested technology that can match the current productivity of the steel industry to greatly reduce carbon emissions. The motivation for developing and implementing one must be based on the urgency of the CO_2 problem, and on appropriate legislation to stimulate change. Modernising production processes with innovative new technologies is an exciting opportunity for the steel industry, but the reduction in CO_2 discharge will make steelmaking more expensive. "Green" steel would, and perhaps should, be more expensive because of the costs associated with creating an improved environment.

What policies are needed to make an overwhelming case for the higher-cost green steel? The partially adopted Paris climate agreement for the protection of the environment is a beginning, but does not represent consensus. The trading of steel across international borders could involve tariffs dependent on the CO_2 pollution associated with its production. The World Trade Organisation permits measures aimed at protecting the environment. Life cycles of steel components can be considered in determining appropriate standards, but ultimately policies must support the green steel concept. Governments can also mandate the use of less-polluting steels in public works programs. Most of the energy needed to produce steel now comes from fossil fuel burning power plants since these fuels are relatively inexpensive. Naturally, anything that results in low-CO_2 electricity will help reduce the CO_2 burden of steel production, albeit indirectly.

7.7   Summary and Outlook

In the last several decades, the steel industry has reduced its energy consumption and carbon dioxide discharge considerably; the energy consumed per tonne of crude steel is now less than half the value in 1960. In addition, recycled steel, which uses just a quarter of the energy required to produce it from ore, constitutes about a third of the annual consumption of steel as an engineered material. Both gradual improvement of energy efficiency and improved recycling have contributed to more sustainable manufacture of steel. However, in view of the data on carbon emission, it has also become clear that it would be impossible to significantly reduce carbon emission without the introduction of radically innovative technology.

There are both technological and commercial barriers to the implementation of new green steelmaking technologies. Numerous ideas have been implemented in the laboratory or pilot plant scales. But actual commitment of these technologies to large scale production will necessitate vast capital investment. An ordinary business case would not justify such investment unless there is a legislation-based motivation for our collective future. Without the new policies, the deployment of green technologies is likely to be in the hopeless future where the global temperature increase has become irreversible.

It is hoped that the growing awareness of the adverse effects of climate change will drive the necessary policy changes needed for the deployment promising green technologies soon. Development and implementation of groundbreaking innovative steelmaking technologies would revitalise the steel industry and at the same time protect our environment.

7.8   Terminology

Blast furnace	A large refractory-lined vertical furnace to produce liquid pig iron containing about 4 wt% carbon. The furnace is charged with iron ore as pellets or sintered chunks, coke, flux, and hot air. A slag consisting primarily of silicates is tapped off periodically for use in road building or other kinds of construction applications. The gaseous exhaust from the furnace contains CO, CO_2 and N_2 is used for heating air or as fuel in the associated plant.
Basic oxygen furnace	A large, pear-shaped reactor for making steel, lined with a basic refractory (calcium-magnesium oxide) that is resistant to attack by slag. The furnace is charged with molten pig iron from the blast furnace, together with steel scrap. A supersonic jet of commercially pure oxygen is then used to oxidise carbon in the pig iron to make steel. The CO produced forms CO_2 at the mouth of the furnace.
Coke oven	Compartments or batteries for heating coal in the absence of air to improve its mechanical strength and remove volatile matter. The exhaust gas from coke ovens is used for heating in various units of steel plants. Combustion of coke provides heat and the reductant CO in the blast furnace.
Electric arc furnace	A furnace for making steel mainly from recycled steel scrap. Large carbon electrodes heat the charge. Oxygen lances are often used for oxidising carbon to CO and CO_2. Oxy-fuel burners are sometimes used for supplemental heating to reduce the consumption of electricity.
Flux	Limestone ($CaCO_3$) or dolomite ($CaCO_3$ and $MgCO_3$) added to the blast furnace to combine with unwanted materials in the charge such as silica (SiO_2) and alumina (Al_2O_3) to make a low melting slag which is a mixture of oxides.
Pig iron:	Iron produced in the blast furnace, rich in carbon $\approx 3.5-4.0$ wt% carbon and smaller amounts of silicon, manganese and other elements. The name "pig iron" originates from the shape of moulds used for casting the iron.
Sintering	Process for agglomerating fine iron ore powder, often with limestone and finely divided coke, by heating with coke-oven gas or natural gas. The heating produces sintered iron ore with good mechanical strength suitable for changing in the blast furnace. This avoids clogging the furnace by permitting the passage of gases between the sintered chunks.
Slag	A mixture of oxides of metals and silicon (to form silicates), and other unwanted materials charged in the blast furnace. Slag is much lighter than liquid iron and liquid slag floats on the metal layer and is periodically withdrawn from the furnace.
Zeolite	A mineral containing silicon, aluminium, sodium and oxygen that can absorb carbon dioxide and other gases.

References

1. Anonymous, Steel's contribution to a low carbon future and climate resilient societies. Tech. Rep., World Steel Association, Brussels, Belgium (2019)
2. Anonymous, World steel in figures. Tech. Rep., World Steel Association, Brussels (2019)
3. D. Carrington, Brutal news: global carbon emissions jump to all-time high in 2018. U.S. edition of Guardian newspaper (2018). https://www.theguardian.com/environment/2018/dec/05/brutal-news-global-carbon-emissions-jump-to-all-time-high-in-2018
4. Y. Junjei, Progress and future of breakthrough low-carbon steelmaking technology (ULCOS) of EU. Int. J. Miner. Process. Extr. Metall. **3**, 15–22 (2018)

Chapter 8
First Bulk Nanostructured Metal

8.1 Introduction

Most people are familiar with crystals; they can be beautiful to look at, sometimes colourful, iridescent and with beguiling symmetry, Fig. 8.1. Engineers also marvel at crystals because they sometimes contain useful defects and at other times attempt to exploit the parts that are coherent in order to tailor the properties of engineering alloys. All of these properties, including the definition of defects, originate from the near perfect order in which the constituent atoms of a crystal are arranged. The crystal is said to have a long-range order, meaning that the entire set of atoms can be described in terms of a small unit of repeating pattern. Therefore, a knowledge of the positions of the handful of atoms in the repeating unit defines the location of the myriads of other atoms in that crystal. It is this order that gives the crystal its shape when allowed to grow without hindrance over a sufficiently long time.

The crystals that we notice with the naked eye are macroscopic, i.e., we can handle them individually and turn them around to perceive their three-dimensional form, without the use of any instruments or aids to visualisation. Each such crystal contains about 10^{23} atoms; to put this into context, the universe contains about 10^{79} particles that have mass. The crystals that we use in engineering can be much larger and have shapes that do not reflect the internal order, but are there to serve a specific role. Figure 8.2a shows a casting of a turbine blade without which we would not have the modern aircraft turbine engine. The blade is grown as a single crystal from the melt, but with the shape consistent with aerodynamic flow rather than the internal symmetry of the arrays of atoms.

We need also to distinguish between the single crystal that has been described thus far and the more usual polycrystalline aggregate of space-filling crystals that form the backbone of most engineering materials (Fig. 8.2b). Obviously, the continuity of long-range periodicity is lost at the junctions between crystals, i.e., the "grain" boundaries. For a polycrystalline material to remain intact following deformation, the change in the shape of one crystal must be matched precisely by

T. DebRoy, H. K. D. H. Bhadeshia, *Innovations in Everyday Engineering Materials*,
https://doi.org/10.1007/978-3-030-57612-7_8

Fig. 8.1 (a) Yellow crystals of sulphur. (b) Crystals of garnet

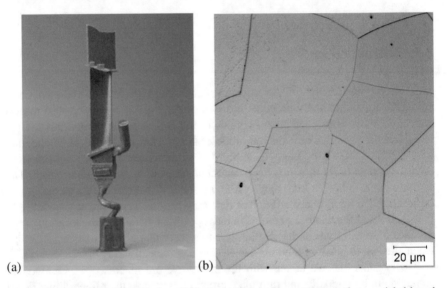

Fig. 8.2 (a) A casting in which the turbine blade is a single crystal made from a nickel based superalloy. (b) A polycrystalline aggregate in a dilute iron–phosphorus alloy (courtesy Jeong Im Kim)

that of its surrounding crystals. This onerous condition propagates throughout the crystal array and results in an increased resistance to deformation, the resistance being greater when the boundaries are closely spaced. Paradoxically, if the deformation is conducted at a sufficiently high temperature, then the boundaries weaken the material because they have a greater free volume when compared against a perfect crystal. Atoms can therefore creep along the boundaries, contributing to unintended strain. This is what limits the life of many rotating components such as some turbine blades. But since most applications are limited to ambient temperatures, major brainpower has been devoted over the last 20 years or so, to the development of polycrystalline structural materials that have a nanostructure, i.e., contain grains that are finer than ever achieved before.

So, how do we define a nanostructure? Nano is an adjective for 10^{-9} — it is sometimes abused — for example, it makes no sense to glorify a small music player or car with this adjective. The term "structure" is better understood by thinking about its opposite, "structureless". A structureless material has uniformity and a traveller within it would soon be bored; in a nanostructure, this dullness would be interrupted every billionth of a reasonable length scale, for example a billionth of a metre (a nanometre) [1]. Imagine then, that there exists a polycrystalline material in which the individual crystals are all a nanometre in size. This would mean that there are, along any direction in that crystal, just four atoms. Such an object does not make sense because it is impossible to define long-range periodicity, and its interface with adjacent nanoparticles would add to the confusion. We therefore have to be reasonable in defining a nanostructure, say a polycrystalline material in which the individual crystals are 20 nm in size. In such a material, there would exist 100 million square metres of interfacial area per m^3. Since strength scales with the inverse of grain size, we should expect impressive properties.

In order to generously exploit a structural material, it needs to be substantial and large enough in all three dimensions to make engineering components. Otherwise, applications are limited to wires or thin sheets. If it is to be used in large quantities then the cost must be affordable, for example comparable to that of bottled water by weight or volume.[1]

Iron is the most stable element in the universe, created in the stars, and available in abundance on the earth. It furthermore is easy to extract from its ores and therefore, affordable in vast quantities — the current production rate is about 1.4 billion tonnes per annum. The question then arises, how do we engineer iron into a nanostructured state?

This is precisely the kind of challenge that a young engineer loves to face, a difficult problem that fires the imagination. The engineer would begin by looking at what has been done in the past, dismissing certain solutions as impossible to scale or unlikely to achieve the goals, identify one or two possibilities and then work in an interdisciplinary team to define a gated process towards the goal. The project may or may not pass through to the highest technology readiness levels, but this would not be for want of trying. We will, in the chapter, follow these principles, beginning with the established methods of increasing strength by refinement.

8.2 Perfection

Back in the 1950s, incredible strength (10 GPa) was obtained in pure iron "whiskers" which are tiny single crystals. This is because in normal materials there are defects known as dislocations that allow small, localised regions of planes of

---

[1]The argument here is that bottled water is not essential and yet we buy it, so therefore it is affordable.

Fig. 8.3 Strength of single crystal whiskers of iron as a function of size [2]

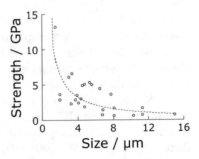

atoms to slide past each other; without them, entire planes would have to translate, making the process of deformation much more difficult. Therefore, crystals that do not contain defects are very strong.

No crystal can be made perfect, no matter how hard we try. This is because there is an elementary principle of thermodynamics that involves a balance between the greater probability of disorder relative to order, and the cost of creating disorder. This ensures that all crystals will at equilibrium contain defects, the concentration of which scales nonlinearly with the number of entities (atoms or molecules) in that crystal. Therefore, the probability of finding a defect in a tiny crystal is much smaller than in one that is large, Fig. 8.3. Therefore, small crystals are stronger but the strength collapses rapidly as the size is increased. This lesson from the 1950s was forgotten when carbon nanotubes came on to the scene with reported values of strength in excess of 130 GPa — it was assumed by many, that this would apply to ropes tens of thousands of kilometres in length. No wonder the space elevator is stuck [3].

The conclusion from the studies of whiskers and the failure to achieve in the case of carbon nanotubes is that reliance on perfection is doomed as the number of entities in the object increases.

8.3 Severe Plastic Deformation

Plastic deformation not only hardens a metal but also refines its structure; ordinary plastic strains in the manufacture of components with complex shapes are limited to about 20% of elongation so the amount of refinement is rather limited. The extent of refinement does not vary linearly with strain. Thus, the effective grain size decreases by a factor of only two when the length of the material is increased by stretching it to seven times its original length. However, the refinement is ginormous if it is stretched to some 8000 times its original length. Indeed, the size of coherent crystalline regions that survive following such severe deformation is a few nanometres, as illustrated in Fig. 8.4, where the sample has a strength in excess of 5 GPa. However, the diameter of the wire itself after such huge extension is just micrometres, finer than the thread used in the manufacture of women's stockings.

Fig. 8.4 An image of the cross-section of a wire, in which each dot represents an atom of iron, and the dark bands are the boundaries between extremely fine crystals. The wire has been made by severe plastic deformation to achieve a strength in excess of 5 GPa [4]

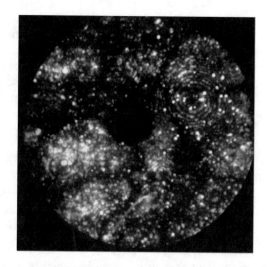

The material does not qualify as "bulk". This is the limitation of severe plastic deformation; there are processes that sequentially bend and unbend the material, resulting in redundant work that leaves the cross-section unchanged. Such processes are ingenious but they involve tortuous processing and are not viable in practice. Severe plastic deformation is not the way forward in the manufacture of *bulk* nanostructured metals.

A difficulty associated with nanostructures in many metallic materials is that they lose the ability to work harden, primarily because the grain interiors are so small and hence do not have a mechanism for work hardening. Work hardening is an essential characteristic to prevent the onset of plastic instability during deformation, where the sample develops a local neck where all the deformation then becomes focused, leading to rapid fracture that can be classified as brittle. Therefore, viable nanostructures require the introduction of alternative work-hardening mechanisms.

8.4 Thermomechanical Processing

One of the greatest achievements of the twentieth century is the invention of microalloyed steel in which minute concentrations of strong carbide forming elements such as niobium or vanadium are added in order to facilitate deformation at very high temperatures where the steel is austenitic, without inducing a coarse austenite grain structure. The carbides, by limiting the growth and recrystallisation of the austenite grains, lead to an even finer ferrite grain structure on transformation during cooling. Fine grains add to the strength of steel, but they also toughen it, i.e., increase its ability to absorb energy prior to fracture.

There are now about tens of billions of tonnes of microalloyed steel, in all sizes and shapes, quietly serving humanity — they are hot-rolled and cooled appropriately to produce a grain size less than a few millionths of metres (μm), with excellent combinations of strength, toughness and weldability. But production of even finer grain sizes requires the transformation to occur at lower temperatures and thermomechanical processing is not capable of doing this when the heat of transformation kicks in and causes recalescence. Therefore, commercial steels of this kind have not succeeded in achieving grain sizes finer than about 1 μm.

8.5 Design Rules for Nanostructures

The discussion above leads to the conclusion that severe deformation, rapid heat treatment, striving for perfection and thermomechanical processing will not lead to viable bulk nanostructured materials. Phase transformation, i.e., generation of fine structures by solid-state reactions is a possibility:

(i) if the transformation can be suppressed to low temperatures to generate the ever finer structures;
(ii) if the rate of transformation can be reduced to avoid recalescence;
(iii) if recalescence can be reduced by storing the enthalpy of transformation within the material;
(iv) if a work-hardening mechanism can be introduced to avoid plastic instabilities.

Diffusion becomes sluggish when transformation temperatures are suppressed, but displacive transformations do not require diffusion. In such a transformation, the choreography of atoms when one crystal structure changes into another determines all aspects of the shape, composition, state, time dependence and properties of the product. In particular, the coordinated dance that the atoms perform, one that is visibly moving to watch, causes upheavals that are better defined as large shear strains. And the upheavals are accompanied by rumbles, the acoustic signals that indicate a relaxation time of just 10^{-5} s. All this occurs in the solid state so the strains force the product crystals to adopt thin plate shapes, a most effective method of grain refinement. Furthermore, these strains are locked within the material and hence consume energy that is otherwise released as the heat of transformation. There are many kinds of displacive transformations, but remember that we desire a slow rate of transformation. A particular reaction known as bainite fulfils all the criteria listed above.

In developing the concept of bainite towards a bulk nanostructured material, scientists reverted to the best theory available to predict that there is no lower limit to the temperature at which bainite can be induced, but that the rate of reaction will diminish dramatically as the steel is forced to transform at temperatures close to ambient. For example, it was estimated that it would take some 100 years to complete the reaction at room temperature. As a compromise, the temperature at which pizza is cooked (200 °C) was selected, taking just 10 days to generate the required structure.

Fig. 8.5
Fe-0.98C-1.46Si-1.89Mn-
0.26Mo-1.26Cr-0.09V wt%,
transformed at 200 °C for 15
days

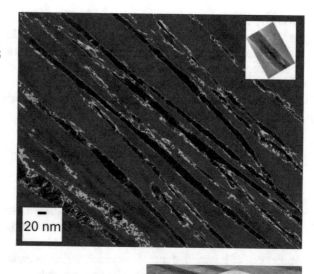

Fig. 8.6 Large-scale manufacture of nanostructured steel: (**a**) shafts for use in future aeroengines, (**b**) plates for the manufacture of superbainitic armour. Photographs courtesy of Rolls Royce plc and Tata Steel, respectively

A steel designed on this basis has been manufactured and characterised [5, 6]; Fig. 8.5 shows the structure obtained following isothermal transformation at 200 °C, consisting of slender platelets of bainitic ferrite with a true thickness in the range 20–40 nm, in an intervening matrix of austenite. The inset on the top right illustrates a carbon nanotube at the same magnification. This austenite is important because when it undergoes deformation-induced martensitic transformation, it enhances the work-hardening capacity of the material, thereby avoiding the usual problem of fine-grained metals where ductility diminishes as the grain size is reduced because the material loses the ability to work harden. Commercially produced examples of the steel are illustrated in Fig. 8.6. Since the original work, many variations of the alloy have been studied and characterised.

The bainite obtained by low-temperature transformation is the hardest ever achieved, with values in excess of 700 HV, a yield strength that can exceed

2000 MPa and a toughness that can exceed 40 MPa m$^{1/2}$.[2] The simple heat treatment involves the austenitisation of a chunk of steel (at say 950 °C), followed by a gentle transfer into an oven at the low temperature (say 200 °C) to be held there for 10 days or so. There is no rapid cooling — residual stresses associated with heat treatment are avoided. The size of the sample can be large because the time taken to reach 200 °C from the austenitisation temperature is much less than that required to initiate bainite. This is an important commercial advantage.

8.6   Concluding Remarks

To summarise, the world's first bulk nanostructured metal is now in production; its essence lies in detailed solid-state phase transformation theory, meticulously researched and argued over many decades. More importantly, there is now a framework of knowledge that allows such materials to be developed systematically, beginning with calculations that define experiments.

Would not it be interesting to see if even finer structures than this can be generated by reducing the transformation temperature even more? Figure 8.7 shows an experiment on a steel that was designed to take 100 years to complete the

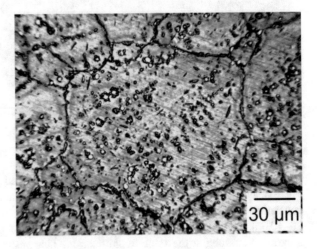

Fig. 8.7 A sample of Fe-1.75C-Si-Mn wt% steel polished perfectly flat and kept in an inert environment at 25 °C at the Science Museum in London. The sample is mainly austenite with some cementite particles; it is the austenite that is expected to transform into incredibly fine bainite at ambient temperature

---

[2]It is simple to make incredibly strong materials but difficult to make them tough at the same time. Toughness is a reflection of the energy that a material can absorb prior to fracture. A high toughness makes the material safer to use. A typical engineering ceramic would have a toughness of just 3 MPa m$^{1/2}$ which is why ceramics cannot be used in components that are subjected to tension.

transformation to bainite. The experiment commenced in the year 2004 and will be completed in 2104. It remains therefore for future generations of scientists to test the theory!

But what does all this mean for life as we know it? The field of nanotechnology has two categories, the functional and structural nanomaterials. Functional nanomaterials focus on small objects of the type used in making devices, whereas the structural variety is after properties on large dimensional scales. In the latter context, there is no other material of the type described in this chapter that achieves the illusive combination of tiny dimensions in huge samples, and at the same time extraordinary mechanical properties that are commercially accessible. Current applications include armour, crushers and fuel injectors [7–10], but it is noticeable that none of these applications requires welding (Chap. 5). The material is almost impossible to join because of its high carbon concentration [11].

8.7 Terminology

Austenite	A crystalline form of iron in which the atoms are arranged at the corners and face-centres of a cubic unit cell.
Bainite	Fine, plate-shaped crystals that grow in austenite when it is appropriately cooled.
Nanostructure	Does not refer to individual crystals, but to a large density of interfaces created in a polycrystalline sample with extremely fine individual crystals, each in a different orientation.
Recalescence	Heat generated within the material when there are changes in the crystals that release latent heat.
Thermomechanical	Thermomechanical processing involves simultaneous deformation and temperature changes imparted on the material being processed.

References

1. H.K.D.H. Bhadeshia, Nanostructured bainite. Proc. R. Soc. London A **466**, 3–18 (2010)
2. S.S. Brenner, Tensile strength of whiskers. J. Appl. Phys. **27**, 1484–1491 (1956)
3. H.K.D.H. Bhadeshia, Large chunks of very strong steel. Mat. Sci. Technol. **21**, 1293–1302 (2005)
4. H.K.D.H. Bhadeshia, H. Harada, High-strength (5 GPa) steel wire: an atom-probe study. Appl. Surf. Sci. **67**, 328–333 (1993)
5. F.G. Caballero, H.K.D.H. Bhadeshia, K.J.A. Mawella, D.G. Jones, P. Brown, Very strong, low-temperature bainite. Mat. Sci. Technol. **18**, 279–284 (2002)
6. C. Garcia-Mateo, F.G. Caballero, H.K.D.H. Bhadeshia, Development of hard bainite. ISIJ Int. **43**, 1238–1243 (2003)
7. H.K.D.H. Bhadeshia, Multiple, simultaneous, martensitic transformations: implications on transformation texture intensities. Mat. Sci. Forum **762**, 9–13 (2013)

8. F.G. Caballero, M.K. Miller, C.G. Mateo, Opening previously impossible avenues for phase transformation in innovative steels by atom probe tomography. Mat. Sci. Technol. **30**, 1034–1039 (2014)
9. C. Garcia-Mateo, T. Sourmail, F.G. Caballero, V. Smanio, M. Kuntz, C. Ziegler, A. Leiro, E. Vuorinen, R. Elvira, T. Teeri, Nanostructured steel industrialisation: plausible reality. Mat. Sci. Technol. **30**, 1071–1078 (2014)
10. A.J. Rose, F. Mohammed, A.W.F. Smith, P.A. Davies, R.D. Clarke, Superbainite: laboratory concept to commercial product. Mat. Sci. Technol. **30**, 1094–1098 (2014)
11. H.K.D.H. Bhadeshia, *Bainite in Steels: Theory and Practice*, 3rd edn. Maney Publishing, Leeds (2015)

Chapter 9
High-Entropy Alloys

9.1 Introduction

Most alloy designs are based, for good reasons, on a solvent-host such as iron, nickel, aluminium, etc. with subtle additions of solutes (alloying elements) that dramatically change their combination of properties for the better. The concentrations required are quite small; for example, a steel containing typically 20 solutes may have a total concentration of just 2 wt%.

During the 1970s, Brian Cantor at the University of Sussex suggested a curious experiment where alloys would be made from equiatomic mixtures of many different elemental components. Although the idea was met with derision, he persuaded a young undergraduate, Alain Vincent, to make two alloys, one containing ten components in equal proportions and the other, twenty. The structure obtained for the twenty-component alloy is illustrated in Fig. 9.1, consisting of many phases. However, one of the phases has mostly Cr, Co, Ni, Fe and Mn, and retained the face-centred cubic crystal structure of nickel. The rest is to some extent history, since we shall see later that precisely this equiatomic, five-component, single-phase alloy has remarkable mechanical properties.

When very large concentrations of solutes are introduced, they may tend to precipitate into crystal structures that are not conducive to good properties. This may be because the precipitates themselves are brittle compounds. In nickel alloys which serve at high temperatures, excessive alloying can lead to the precipitation of the so-called topologically close-packed structures which often lead to a decrease in the ability of the material to resist fracture. Alloys with large concentrations of solute also suffer from chemical segregation during production on a large scale.

A few of these difficulties can be mitigated if the alloying elements can be persuaded to stay mixed at an atomic level in spite of their large concentrations. What then can persuade diverse atoms to stay mixed rather than separate?

Figure 9.2a shows a cylinder containing an ideal gas in which the atoms have moved in such a way that they are all located on one half even though there is

© Springer Nature Switzerland AG 2021
T. DebRoy, H. K. D. H. Bhadeshia, *Innovations in Everyday Engineering Materials*,
https://doi.org/10.1007/978-3-030-57612-7_9

Fig. 9.1 The original
twenty-component alloy (Mn,
Cr, Fe, Co, Ni, Cu, Ag, W,
Mo, Nb, Al, Cd, Sn, Pb, Bi,
Zn, Ge, Si, Sb, Mg.), melted
in a crucible using pure
components in equiatomic
proportions, and air cooled.
Image courtesy of Brian
Cantor [1]

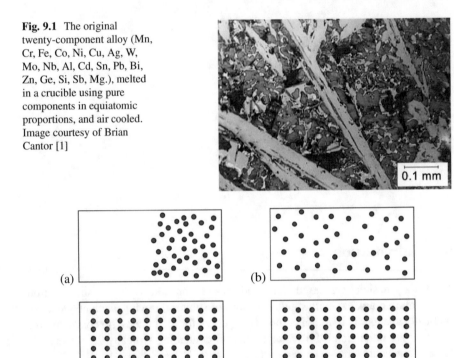

Fig. 9.2 (**a**) A cylinder where all the atoms of an ideal gas are located on one half. (**b**) The same
cylinder but with atoms uniformly distributed. (**c**) A crystal in which the two halves have different
atoms, blue (A) and red (B). (**d**) The same crystal where the atoms are mixed

no barrier within the cylinder. The corresponding scenario in Fig. 9.2b where the
atoms are more uniformly dispersed within the cylinder, is, as common sense would
indicate, much more *probable*. The first case is more organised and the second more
disorderly.

Figure 9.2c shows a crystal in which like atoms are separated into the two halves.
There is therefore, only one *configuration* in which the atoms can be arranged.
Figure 9.2d is the case where they are more homogeneously distributed over the
crystal. The number of ways in which the atoms can be arranged for this second
scenario is far greater. Suppose there are N sites among which we distribute n atoms
of type A and $N - n$ of type B. The first A atom can be placed in N different ways
and the second in $N - 1$ different ways. These two atoms cannot be distinguished
so the number of different ways of placing the first two A atoms is $N(N - 1)/2$. If
we proceed to place all the atoms in this way, the number of distinguishable ways
of placing all the A atoms is

$$\frac{N(N-1)\ldots(N-n+2)(N-n+1)}{n!} = \frac{N!}{n!(N-n)!}. \tag{9.1}$$

So if the atoms behave ideally, i.e., they do not have a preference for the type of neighbour, then the probability of an even distribution is much more likely than the ordered distribution.

For a real system for which the number of atoms is very large, a parameter is needed that expresses the likelihood as a function of the correspondingly large number of configurations (w) possible. Suppose that a term S is defined such that $S \propto \ln w$, where the logarithm is taken because it may be necessary to add two different kinds of disorder (after Boltzmann), then the S is identified as the configurational *entropy* $S = k \ln w$, where k, the proportionality constant, is known as the Boltzmann constant which for a mole of atoms is the gas constant R. The entropy is a thermodynamic function of state and it is additive. When comparing scenarios, the one that is favoured on the basis of the degree of disorder is that which has the greater entropy. In terms of solutions, entropy favours mixing over separation. On this basis, it can be shown quite simply that the change in entropy when atoms mix is given by

$$\Delta S = -R\, \Sigma_{i=1}^{j} x_i \ln x_i$$

where $i = 1 \ldots j$ represents the atomic species and x_i its mole fraction; R is the gas constant.

This simple equation proves that maximising the kinds of atoms in solution makes the solution more likely assuming that the atoms are indifferent to their neighbours. For say a five-component system where the concentration of each species of atoms is identical, $\Delta S_M = 1.61R$, and diminishes to $0.69R$ if there are just two components in equal concentration.

This is the basis of the so-called high-entropy alloys consisting of numerous components in approximately equal concentrations [2]. By doing so, either a single-phase solution is obtained or the number of phases is minimised. For example, equiatomic CrMnFeCoNi is a single face-centred cubic solid solution with an impressive combination of strength and toughness, even at cryogenic temperatures [3].

Figure 9.3 shows the entropy of mixing for three components; as expected, the largest values are obtained when the concentrations of each of the constituents are equal. The constituents have been identified as A, B, C but could be replaced by any element without changing the plot as long as the solutions thus formed are *ideal*. This means that the atoms are dispersed at random, they see no difference in bonding irrespective of their neighbouring atoms. A random distribution of atoms is not possible for real alloys which never are ideal solutions (although very high temperatures can randomise the distribution). This complication is discussed next.

Fig. 9.3 Entropy of mixing of ternary alloy showing that an equimolar alloy has the maximum entropy. The numbers on the equal entropy contours are in J mol^{-1}. The dilute alloys are shown near the three corners while the concentrated ones are near the middle

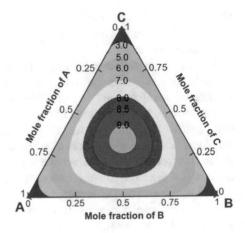

9.2   A Complication

That entropy is identified as a factor that is favourable to the mixing of atoms is clear. Temperature itself is defined as the inverse of the rate of change of entropy with the internal energy, so the contribution of the configurational entropy to the energy of system is $-T\Delta S_M$.

However, the nature of the bonding between atoms has thus far been neglected. To form a solution of A and B atoms requires the breaking of A-A and BB bonds to generate some A-B bonds. The resulting change in energy is expressed as an *enthalpy* of mixing, ΔH_M. A negative value of this quantity implies that A atoms prefer to have B near neighbours, whereas the clustering of like atoms is favoured if it is positive. Therefore, the enthalpy change, depending on its sign, can either favour or oppose the formation of a solution at low temperatures. The net change due to both enthalpy and entropy is called a free energy, with

$$\Delta G_M = \Delta H_M - T\Delta S_M.$$

If ΔG_M is positive, then the mixing of atoms cannot occur spontaneously.

Note, however, that ΔG_M is not a guide to the stability of solution. The concept of stability cannot be considered in isolation for a single solution; stability is only defined with respect to something else. For example, with respect to the free energy of a precipitate, or the decomposition of the solution into composition-rich and composition-depleted regions.

There exist no alloys for which $\Delta H_M = 0$ so concepts connected with high-entropy alloys cannot simply rely on the entropy of mixing. If it is not possible to calculate ΔH_M and other thermodynamic quantities relevant to determining the relative stability of the solution, then alloy design must be based on empirical rules such as those proposed by Hume-Rothery almost a century ago [4]:

Fig. 9.4 Plot of the magnitude of the entropy contribution to that of enthalpy, versus a measure of the atomic misfit in the multicomponent solution. x_i and r_i are the mole fraction and atomic radius of species i, and $\bar{r}$ is the mean atomic radius. Data from Zhang et al. [6]

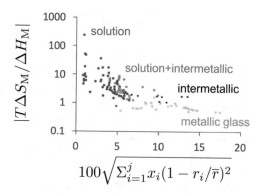

$$100\sqrt{\Sigma_{i=1}^{j} x_i(1 - r_i/\bar{r})^2}$$

1. Unless the solute and solvent differ atomic size by less than about 15%, they cannot form extensive solid solutions.
2. The electrons in a solid are to some extent shared between atoms. An atom's ability to attract these electrons is defined as electronegativity. A large difference in electronegativity when compared with the host implies that the atoms are more likely to form a compound.
3. Valency is defined as the combining power of an atom. A metal with a lower valency is more likely to dissolve in one that has a higher valency, than vice versa.

The rules which were proposed at a time when the metallic state was still regarded as a problem [5], but today are applied widely in the qualitative design of high-entropy alloys, with some success.

Figure 9.4 shows a plot of the "strength" of the mixing entropy relative to the enthalpy of mixing, against the atomic misfit in the multicomponent solution. It is evident from the experimental data that a low misfit and large ratio favour the formation of a single solid solution in high-entropy alloys, if that is what is desired.

An alternative approach is first-principles calculations that use electron theory to reveal energies of particular structures. The limitations there are that most such calculations pertain to zero Kelvin and zero pressure, and rarely are able to deal with a random solid solution.

In all of these methods, it is important to realise that solubility cannot be considered in isolation for a given solution. For example, it is not proper to state that ΔG_M defines the stability of the solution. The limits to solubility depend on equilibrium between two different phases; a simple example illustrates this: the maximum amount of salt that can be dissolved in water is different from the corresponding limit to the solubility of sugar in water.

9.3   A Particular High-Entropy Alloy

Figure 9.5 shows the crystal structures of the constituents of a high-entropy alloy, CrMnFeCoNi containing equal amounts of each of the elements listed [7]. An interesting outcome is that although the crystal structures are quite diverse, including the fact that the unit cell of Mn has 58 atoms, they all adopt the nickel face-centred cubic structure. This outcome is good because this structure in other circumstances is known to be nice and ductile even at low temperatures.

Qualitatively, the structure of cobalt differs little from that on nickel — both can be considered as layers of close-packed atoms, albeit in different stacking sequences. And cobalt transforms into a face-centred cubic structure at temperatures above $\approx 420\,^\circ$C. Iron although body-centred cubic, also can exist routinely in the face-centred cubic structure. Chromium and manganese are more difficult to rationalise in this way, but the fact remains that the high-entropy alloy containing all the constituents has a face-centred cubic structure.

Figure 9.6 shows the microstructure of CrMnFeCoNi consisting of equiaxed grains containing parallel-sided features that are known as annealing twins that normally characterise the microstructure of pure nickel. The structure illustrated is not unusual but it is what happens during deformation that really gives the alloy its properties, as described below.

Fracture toughness is a material parameter that permits the calculation, for a given stress, of a crack length beyond which the material fails by rapid crack growth. The toughness is one of the most important properties in ensuring the safe design of engineering structures. The yield strength represents the stress at which the material begins to deform in an irreversible manner. A material in a structure should not in general be loaded to beyond the yield strength. Figure 9.7 shows a plot of the properties of CrMnFeCoNi compared against other materials. The combination of

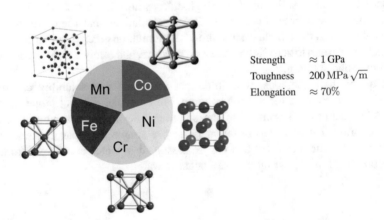

Strength	$\approx 1\,$GPa
Toughness	$200\,$MPa $\sqrt{m}$
Elongation	$\approx 70\%$

Fig. 9.5 High-entropy alloy containing equal atomic fractions of Cr, Mn, Fe, Co,Ni. The crystal structures of the elements at ambient temperature and pressure are illustrated; the alloy adopts the face-centred cubic crystal structure of nickel

Fig. 9.6 Microstructure of
single-phase CrMnFeCoNi.
Micrograph courtesy of
Bernd Gludovatz, Eso George
and Robert Ritchie

Fig. 9.7 A plot showing the
fracture toughness against the
yield strength for common
materials and for the
CrMnFeCoNi alloy (red
dots). Adapted using data
from Gludovatz et al. [3]

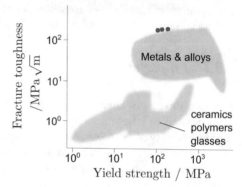

toughness and strength is impressive though when it comes to experimental alloys
such as CrMnFeCoNi, it should be borne in mind that equally if not more impressive
properties are found in much simpler metallic alloys [8].

9.3.1   Making High-Entropy Alloys

A high-entropy alloy contains a mixture of large concentrations of a variety of
elements. It can be produced by melting and casting if the melting temperatures of
the constituent metals are not very different. But the difference that can be tolerated
is large because it is often the case that an alloy has a lower melting temperature
than the constituents, as listed below:

Cr	1907 °C	Co	1495 °C
Fe	1538 °C	Mn	1246 °C
Ni	1453 °C	**CrMnFeCoNi**	1284 °C

There is no commercial production of large quantities of high-entropy alloys so the
most common method used for melting is an electric arc furnace, although electrical

resistance heating or induction heating have also been used to produce small melts (kg). This would be done in a controlled atmosphere or vacuum to avoid oxidation.

Some alloying elements are highly reactive or have melting points that are significantly different. For these systems, it is difficult to make a homogeneous alloy. One example is AlLiMgScTi where several of the alloying elements are reactive and the melting points of the constituent elements differ by well over 1000 K. In such cases, severe deformation in the solid state can force the elements into solution but the process is complex and the final powder would require consolidation.

The phases present in an alloy are often influenced by the post production cooling rate. For systems where several phases are anticipated under equilibrium conditions, not all the equilibrium phases may actually form because of kinetic limitations. Such limitations occur if the diffusion is slow or if there is insufficient time for a phase to form. When an equiatomic AlCoCrCuFeNi alloy is cooled rapidly it becomes a body-centred cubic single-phase whereas slow cooling causes the appearance of two additional face-centred cubic phases.

9.3.2   Properties

Having a mixture of atoms in equal concentration slows the diffusion rate; this has been mooted as a benefit to creep deformation but account must be taken of the thermal stability of the solution since precipitation may occur. The different sizes of atoms contribute to solid solution strengthening. In fact the strength and hardness levels are similar to those of conventional alloys with ductility decreasing as the strength is increased, although there are clearly exceptions such as the CrMnFeCoNi alloy where ductility actually increases with strength because of microscopic twinning during deformation, that delays the plastic instability responsible for final fracture.

Although large numbers of high-entropy alloys have been studied and properties measured [6], proper structural integrity measurements in terms of fracture toughness and tensile tests are few. For example, the very high strengths (2–3 GPa) reported for $AlCoCrFeNiTi_x$ alloys are based on compression experiments which are not suitable for an assessment of a structural material that is likely to be loaded in tension. The samples involved are usually too small to result in a reliable assessment of the influence of defects.

It is not surprising that there are perhaps a handful of applications of high-entropy alloys. Of the nine applications listed in a recent review, only one is apparently used, as a brazing alloy for joining ceramics to metals or dissimilar metals. All the others are promises.

9.4   Concluding Remarks

The science of high-entropy alloys might be considered to be mature in the sense that the priorities might be better focused on innovations in applications. Such a focus may reveal new science in that an application requires the consideration of a bank of properties measured on realistic samples. The properties obtained will need to be exceptionally good if the expense of the alloying elements is to be acceptable. The scaling up of production is likely to pose enormous challenges to avoid gross chemical segregation.

9.5   Terminology

Alloy	A metallic material made up of two or more elements.
Configurational entropy	Entropy contribution resulting from the position of the atoms. It does not consider the velocity of the atoms.
Creep	Time dependent deformation of metallic materials at elevated temperatures under load.
Crystal	A solid with atoms positioned in repeatable three-dimensional arrays. Metals are crystalline.
Enthalpy of mixing	The heat absorbed or evolved due to mixing. The enthalpy of mixing is positive for endothermic reactions that require absorption of heat and negative for exothermic reactions that produce heat due to mixing.
Entropy	Lack of order, i.e., the degree of disorder or randomness in a system. Its value depends on variables such as composition, temperature and pressure.
Grains	Crystals within a solid separated by clearly demarcated boundaries.
Ideal solution/mixing	It is a solution of elements with no energy change. The volume of the solution is the sum of the volumes of all constituents. In thermodynamics, it is often defined as a solution in which the activity of an element equals its mole fraction.
Injection moulding	A process of manufacturing parts by injecting alloy powders mixed with a binder into a mould cavity to produce many small parts quickly. After moulding, the parts are heated to remove the binder and sintered to gain mechanical strength.
Lattice	Three-dimensional array of points where atoms are located in a crystalline solid.
Laves Phase	Intermetallic compound with a formula of AB_2 where A and B are two metals. Their crystal structure may be cubic or hexagonal. Precipitates of Laves phases have been reported in HEAs.
Molar concentration	It indicates concentration of an element or its amount per unit volume. The amount is expressed in moles, i.e., weight divided by atomic weight.
Ordered alloy	An alloy where the positions of the atoms alternate with some regularity.
Random mixing	Mixing of atoms without any order. The locations of atoms are unpredictable in an alloy where the atoms are randomly mixed.
Strength	Ability to resist deformation under load.
Superalloys	Nickel, cobalt or iron alloys with excellent properties such as strength, oxidation and corrosion resistance at elevated temperatures.
Toughness	Ability to absorb energy before failing under load.

References

1. A.J.B. Vincent, A study of three multicomponent-alloys. Master's Thesis, University of Sussex, Part II thesis, Sussex (1981)
2. K.H. Huang, J.W. Yeh (advisor), A study on the multicomponent alloy systems containing equal-mole elements. Master's Thesis, National Tsing Hua University, Taiwan (1996)
3. B. Gludovatz, A. Hohenwarter, D. Catoor, E.H. Chang, E.P. George, R.O. Ritchie, A fracture-resistant high-entropy alloy for cryogenic applications. Science **345**, 1153–1158 (2014)
4. W. Hume-Rothery, Researches on the nature, properties and conditions of formation of intermetallic compounds, with special reference to certain compounds of tin. J. Inst. Metals **35**, 295–261 (1926)
5. J.D. Bernal, The problem of the metallic state. Trans. Faraday Soc. **25**, 367–379 (1929)
6. Y. Zhang, T.T. Zuo, Z. Tang, M.C. Gao, K.A. Dahmen, P.K. Liaw, Z.P. Lu, Microstructures and properties of high-entropy alloys. Progr. Mater. Sci. **61**, 1–93 (2014)
7. B. Cantor, I.T.H. Chang, P. Knight, A.J.B. Vincent, Microstructural development in equiatomic multicomponent alloys. Mater. Sci. Eng. A **375–377**, 213–218 (2004)
8. J. Park, K. Lee, H. Sung, Y.J. Kim, S.K. Kim, S. Kim, Fracture toughness of high-Mn steels at room and cryogenic temperatures. Metall. Mater. Trans. A **50**, 2678–2689 (2019)

Chapter 10
Metals That Do Not Forget

10.1 Introduction

In a crystalline material, the pattern in which the atoms are arranged can be changed in one of two ways. The bonds could all be broken and the atoms rearranged into a new structure without regard to their location in original crystal. Such a reconstruction of the structure loses all correspondence between the atoms of the two structures; this process can happen only when the temperature is high enough to permit the diffusion of atoms.

There is, however, another mechanism by which structural change, i.e., the transformation, can be achieved without diffusion. Figure 10.1 shows a crystal containing two kinds of atoms. Before transformation, the defining unit cell of the pattern is identified as the blue square. To achieve transformation, the crystal is sheared homogeneously to generate the new pattern of atoms, identified by the oblique, green unit cell. The important distinction here is that the sequence in which the atoms are arranged remains identical before and after transformation. This means that if the transformation is reversed, then the original crystal is recovered in an exactly identical configuration. It is entirely accurate to describe this as a *memory* because a clear atomic correspondence is maintained between the parent and product crystals.

A transformation like this is said to be *martensitic*. The name has its origins in the early metallographic work on steels by Osmond, who in 1868 published a paper in which he named the constituent as martensite, after Adolf Martens who was one of the pioneers in the microscopy of metals. Figure 10.1 shows another vital characteristic of martensitic transformations. If the unit cell of the parent is changed into a different shape, then the macroscopic shape of the entire crystal must also change. This *shape deformation* can be observed experimentally (Fig. 10.2). When characterised in detail, the deformation is a large shear and a small volume change. The transformation therefore is not simply a change in crystal structure but a physical deformation that can, like any other deformation, do work.

© Springer Nature Switzerland AG 2021
T. DebRoy, H. K. D. H. Bhadeshia, *Innovations in Everyday Engineering Materials*,
https://doi.org/10.1007/978-3-030-57612-7_10

Fig. 10.1 The crystal on the left contains two kinds of atoms arranged in a particular pattern identified by the blue cell. The upper part of the crystal is homogeneously deformed to generate the new pattern illustrated on the right

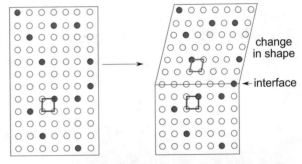

Fig. 10.2 The upheavals caused by martensitic transformation. The parent phase was polished flat before transformation in this Fe-0.31C-30.5Ni wt% alloy. After Forsik and Bhadeshia

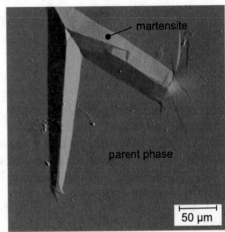

10.2   Shape Memory

Martensitic transformation can be induced by an external stress, subjecting the material to temperatures where the martensite is more stable than the parent phase, or using magnetic fields. The *reversibility* of the parent⇌martensite change requires the interface, i.e., the boundary separating the two phases, to move without creating crystal defects. Not all martensitic transformations are easily reversible because numerous defects are created when the martensite forms. This is particularly so when the plates grow at high temperatures where the phases are relatively weak and hence unable to elastically accommodate the large shape deformation accompanying transformation.

Figure 10.3 shows that the displacements caused by martensite that has formed during the cooling of a Cu-Al-Ni-Mn-Fe alloy disappearing as the sample is heated over a narrow temperature range, to regenerate the parent phase. This is known as a thermoelastic martensite, highly reversible, a shape memory metal because the overall shape can change between that of martensite and austenite. A consequence of the atomic correspondence mentioned in the introductory remarks, and the fact that few defects that would interfere with reversibility, are created during the repeated

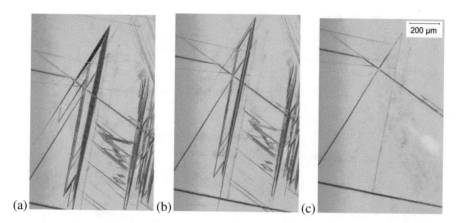

Fig. 10.3 Martensite shape deformation being reversed in a thermoelastic Cu-Al-Ni-Mn-Fe alloy that is heated (**a**) At 301 K. (**b**) At 302 K. (**c**) At 303 K. The micrographs are snapshots from a movie kindly provided by M. Benke, V. Mertinger, Institute of Physical Metallurgy, Metalforming and Nanotechnology, University of Miskolc

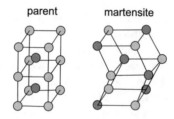

Fig. 10.4 The primitive cubic structure of the parent phase with a Ni atom at the centre of the cube and there are two cubes stacked on each other for clarity. The martensite structure is monoclinic, i.e., none of its cell edge is equal and one of the angles deviates from 90 °C

motion of the interfaces. This effect was discovered by Khandros and Kurdjumov in 1949 [1] using the same sort of technique as illustrated in Fig. 10.3, on a Cu-Al-Ni alloy.

Perhaps the most reversible shape memory alloy was discovered in 1959 by Buehler [2]. It consists of an almost equiatomic mixture of nickel and titanium as illustrated in Fig. 10.4.

10.3   Applications

The alloy has many applications, for example in elastically bendable spectacle frames, actuators on a specific part of an aircraft engine, jewellery, etc. But perhaps, the one that has had the most impact is the stent that is inserted remotely into a blood vessel that is close to being blocked by deposits. The large recoverable strains

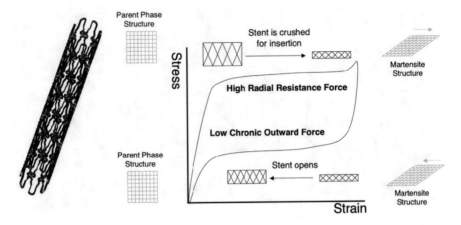

Fig. 10.5 On the left is a model of a stent cut from a nitinol tube using a laser. The diagram shows the process by which the stent is delivered to the location within the human body. After Morgan [3], with permission of Elsevier

and hysteresis in the stress–strain behaviour during transformation make Nitinol particularly suitable for this purpose. The materials has, for specific applications, been approved for insertion into the human body.

The stent is made by laser machining Nitinol tube into a pattern that makes it flexible. Figure 10.5 shows how the stress–strain behaviour exhibits a hysteresis during reversal of transformation. The initial large diameter machined-tube which represents the *open condition* is in the parent crystal structure. It then is compressed into a diameter for insertion into a delivery catheter, whence it has a martensite structure. On reaching the selected delivery location, the tube is pushed out and expands back on encountering the body-temperature into the parent structure, the expansion facilitating the opening of the blood vessel. The force required for this reversal is small, and importantly, almost constant so there is not excessive loading on the vessel itself. There are countless people who have benefitted from this technology.

Nitinol type alloys have been used in art forms, for example a large outdoor-sculpture that changes its shape with the weather. The large earring illustrated in Fig. 10.6a has the ability to recover its shape if distorted during an embrace. There are many small devices that are actually in production for use in automobiles; Fig. 10.6b shows a thermostatic valve where the shape memory element is the shiny spring which with temperature change, operates with sufficient force to switch from the on to the off position.

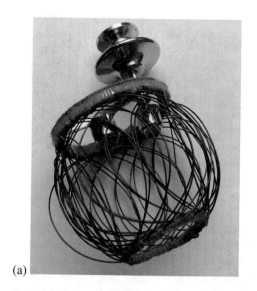

(a)   (b)

Fig. 10.6 (a) An earring which uses a NiTi shape memory alloy for the basket. As a result, the basket recovers its shape if deformed, for example, during an embrace. The earring was a collaboration between Professor Jan van Humbeek and an artist. (b) A thermostatic valve using a spring made of shape memory metal. Picture courtesy of Burkhard Masss (Ingpuls GmbH), and Gunther Eggler

10.4 Rubber-Like

It is worth repeating that martensitic transformation produces a deformation. Because the crystal symmetry is lower than that of the parent when martensite forms, a large number of differently oriented crystals of martensite can grow in a single crystal of the parent. Each one of these will deform in a different direction so on average, the net deformation over the whole of the original crystal may be small. If this polycrystalline aggregate of martensite is now aged to allow very subtle atomic movements that stabilise the structure, then it makes it difficult to revert easily back into the parent structure. When stressed, that martensite plate which best accommodates the stress will grow and consume the other martensite crystals, until a single crystal of martensite is obtained, so the net deformation now becomes large. On the removal of the stress, the single crystal reverts back to the original polycrystalline state, thus reversing the deformation, Fig. 10.7.

Essentially the back and forth conversion of many martensite plates into a single one results in large recoverable strains that are analogous to the stretching and unstretching of polymeric rubber.

There are many other phenomena associated with the shape memory effect; these are too detailed for the present purposes but are reviewed in [4].

Fig. 10.7 Schematic
illustration of how a
polycrystalline cluster of
martensite changes into a
single crystal under the
influence of stress, only to
revert to the original state if
the stress is removed

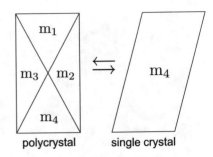

10.5   Loss of Memory

When interfaces move, whether they are between the parent and martensite or between different orientations of martensite, there is some loss of perfection in the respective crystal structures. Anything that disrupts the long-range order of a crystal is a defect. Such defects accumulate with the number of cycles of reversal. They eventually reach sufficient number density to clog the interface and stop it from moving. The shape memory effect is then lost. The loss can occur over a million cycles in the case of Nitinol, or a few cycles in the case of iron-based alloys. Some applications only require the reversal once; an example is a sleeve for joining dissimilar metal pipes. The sleeve is cryogenically cooled and then expanded mechanically. After placing it in position around the abutting pipes, warming induces it to reverse via the shape memory effect, into a contracted form which firmly grips the pipes; the so-called shrink-to-fit pipe couplers. One example of a shape memory alloy suitable for this purpose is Fe-14Mn-6Si-9Cr-5Ni wt% because it also exhibits good creep and stress–relaxation behaviour consistent with the reliable operation of the joint [5].

There also are limits to the amount of strain that should be applied. This is because any changes induced by applying stress must only be accommodated by inducing corresponding changes in crystal structure or crystal orientations. They must not cause irreversible plastic deformation.

10.6   Concluding Remarks

The shape memory effect in metals fires the imagination and has made enormous contributions to real life, including life-saving or life-extending medical interventions. And there are many other medical applications that have been reviewed elsewhere [3].

Although there are many alloys that exhibit the effect, Nitinol has made the biggest impression because of its exceptional reversibility. According to historical accounts, Buehler made two melts, one of which he dropped intentionally to hear a bell-like ring indicating the quality he perceived was needed for a particular application. The other one was dropped after cooling and gave a dull thud. This was taken to mean that the alloy had a double state, and the rest is history. But it emphasises the sound-deadening ability of the alloy due to changes in the crystals within; this in itself is a quality that is now exploited industrially.

Current alloys exhibiting shape memory effects are limited to applications below about 200 °C. There is a need for alloys that exhibit the shape memory effect at elevated temperatures for use, for example, in high-temperature engines. Some work has been done on tantalum–ruthenium and niobium–ruthenium alloys [6], which do indeed show shape recovery but the recoverable strain is small and there are likely to be effects due to diffusion at high temperatures.

10.7 Terminology

Homogeneous deformation	A deformation that is uniform, so that points that are initially colinear remain so, and lines that are initially coplanar remain so after the deformation.
Martensite	A crystal that grows from the parent phase without diffusion, accompanied by a change in the shape of the transformed region.
Shape deformation	The change in the shape of a parent phase when it transforms into martensite.
Shape memory	An effect in which a material that is set into a specific shape and deformed into another, can be stimulated to recover its original shape, often repeatedly.

References

1. G. Kurdjumov, L.G. Khandros, First reports of the thermoelastic behaviour of the martensitic phase of Au-Cd alloys. Dokl. Aka. Nauk SSSR **66**, 211–213 (1949)
2. G.B. Kauffman, I. Mayo, The story of Nitinol: the serendipitous discovery of the memory metal and its applications. Chem. History **2**, 1–21 (1997)
3. N.B. Morgan, Medical shape memory alloy applications - the market and its products. Mater. Sci. Eng. A **378**, 16–23 (2004)
4. H.K.D.H. Bhadeshia, C.M. Wayman, Phase transformations: nondiffusive, in *Physical Metallurgy*, ed. by D.E. Laughlin, K. Hono, chap. 9. (Elsevier, North-Holland, 2014), pp. 1021–1072
5. J.C. Li, X.X. Lu, Q. Jiang, Shape memory effects in an Fe14Mn6Si9Cr5Ni alloy for joining pipe. ISIJ Int. **40**, 1124–1126 (2000)
6. R.W. Fonda, H.N. Jones, R.A. Vandermeer, The shape memory effect in equiatomic TaRu and NbRu alloys. Scr. Mater. **39**, 1031–1037 (1998)

Chapter 11
Low-Density Steels

11.1 Introduction

Density features in many of the criteria used in the selection of materials for specific engineering applications. Sometimes a large density is advantageous if inertia becomes important or for the storage of energy, but for other applications such as transportation of all kinds, a low-density can contribute to energy efficiency.

Consider first the case where a high density helps. Internal combustion engines work using pulses of power generated momentarily when fuel is ignited in a cylinder. The resulting pulse of pressure drives mechanical motion, but the system on its own would lead to jerky movement. The process can be smoothed by connecting a flywheel that stores kinetic energy, with the latter scaling with the mass of the wheel. Steel is an eminently appropriate choice for the flywheel because it has a relatively large density and other desirable properties that ensure structural integrity at an affordable cost.

Some applications require dead weight, in which case a dense material can be an advantage. The tallest of skyscrapers can sway at the top by a metre under elastic deflection due to gusts of wind or earthquakes; this can cause discomfort, shock and nausea to the occupants. In the Taipei 101 building, such deflections are compensated for by a tuned-mass damper which consists of a steel ball weighing 660 tonnes, suspended using steel ropes, from the 91st floor down to the 87th floor of the building (Fig. 11.1). The giant ball is assembled from steel slabs which then are welded together. Any swaying of the building causes this pendulum to move in the opposite direction. The ball is connected to eight viscous dampers that absorb the kinetic energy during building excitation, thus greatly mitigating the motion of the building. The natural frequency of the building under the action of external stimuli determines the designed natural frequency of the damping system, hence the term "tuned". The steel ball in the Taipei building consists of slabs that are cut such that on assembly they approximately form a sphere and the whole damping system is painted to be aesthetically pleasing as a tourist attraction. The volume

T. DebRoy, H. K. D. H. Bhadeshia, *Innovations in Everyday Engineering Materials*, https://doi.org/10.1007/978-3-030-57612-7_11

Fig. 11.1 (**a**) Tuned-mass damper weighing in excess of 660 tonnes showing the viscous dampers attached to the lower half of the ball assembly. (**b**) The steel cables that cradle and suspend the massive ball from the 91st storey of the Taipei 101 tower, which is 509.2 m tall. Photograph authorised by Jessie Lin of TAIPE 101, via the good offices of Professor J. R. Yang

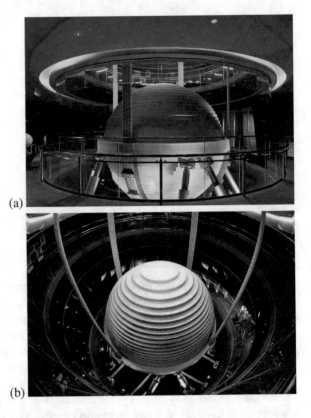

(a)

(b)

of the equivalent magnesium would be roughly five times greater, and an order of magnitude more expensive; the cradle and viscous dampers would remain based on steel.

11.2   Range

The amount of matter that is packed into a unit volume has a huge range in nature, Table 11.1. The data are mostly presented under standard temperature and pressure conditions in order to permit comparisons. The decompressed density of the Earth is calculated from its average density ($>5\,\mathrm{g\,cm^{-3}}$) so that pressure variations across its depth are eliminated. The neutron star is the gravitationally collapsed core of a star that stopped generating heat, the collapse essentially eliminating the free space within atoms to create a huge cluster of tightly packed neutrons. What the star illustrates in particular is that ordinary atoms as we perceive them on earth, consist mostly of empty space, and even when packed together in solids, can be compressible. The packing itself can be disturbed by introducing foreign atoms that have different size and electronic structure. For example, placing an aluminium atom

Table 11.1 Densities of some substances *measured* at standard temperature and pressure (STP), i.e., 0 °C, 100 kPa, unless otherwise indicated. There are other elements, synthetic elements such as hassium ($\approx$41 g cm^{-3}), which are *expected* to be denser than osmium but the density of which has not been measured

Substance	Structure	Density/g cm^{-3}	Conditions	Reference
Densest element, osmium	Hexagonal close-packed	22.589	STP	[1]
Densest liquid, mercury	–	13.54	STP	[2]
Densest gas, radon	Gaseous	0.00973	STP	[3]
Decompressed earth	–	4.0–4.5	STP	[4]
Neutron star	–	10^{14}	–	[5]
Iron	Body-centred cubic	7.874	STP	[6]
Magnesium	Hexagonal close-packed	1.737	STP	[7]
Aluminium	Cubic close-packed	2.708	STP	[8]
Silicon	Diamond cubic	2.329	STP	[9]

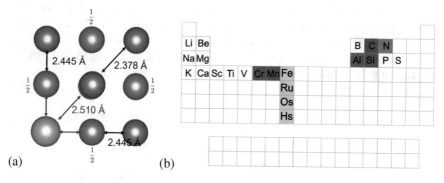

(a) (b)

Fig. 11.2 (**a**) Change in interatomic spacing caused by the introduction of an aluminium atom (blue) in the face-centred cubic form of iron. The diagram is a projection of the cell normal to a cube edge. The fractions indicate the height of the atom in the direction normal to the diagram; unlabelled atoms are at heights 0,1 [10]. (**b**) Periodic table showing the chemical analogues of iron (blue), and elements that have a smaller atomic mass than iron. Elements that are gaseous at room temperature and pressure are excluded

in face-centred cubic iron causes an expansion in the space around the aluminium atom (Fig. 11.2a), which we shall see later leads to an overall decrease in density.

While iron is regarded in general to be a dense metal, it is the iron analogues in the periodic table that break the records for density. In Fig. 11.2b, the blue column represents these analogues, so-called because they have similar outer electron structures, and hence chemical activity. Density increases down the column, with osmium being the most dense element that has ever been characterised experimentally. Hassium, if it could be synthesised in sufficient quantity and if it were to be stable for a sufficiently long time, might be measured to have a density roughly double that of osmium.

11.3 Manipulation of Density

The simplest way to reduce the density of steel is to add solutes that either
increase the lattice parameter of pure iron, or reduce the average atomic mass of
the alloy, or both. Figure 11.3 shows how aluminium not only increases the lattice
parameters of both ferrite and austenite, which combined with its smaller atomic
mass, significantly reduces the net density of the alloyed iron.

Figure 11.2b shows elements that are lighter than iron, excluding gases other
than nitrogen. Those elements highlighted in pink can in practice be added to steel
in significant quantities, and there are circumstances where they would benefit a
range of properties other than density. The earliest work focused on reduced-density
steels comes from Russia [13, 14], although aluminium has long been considered
as a solute that enhances the oxidation resistance of steel [15]. Aluminium does
reduce the stability of austenite, which is essential when the microstructure requires
to be generated by its solid-state transformation. Low-density alloys therefore
contain significant concentrations of manganese and carbon to balance the effect
of aluminium by increasing the thermodynamic stability of austenite. The Russian
low-density steels contained typically Fe-25Mn-10Al-1C wt%; it is notable that Mn
and C, being lighter elements than iron, both lead to small but significant reductions
in density, so that the Fe-Mn-Al-C alloy ends up with a density of about $6.8 \, \mathrm{g \, cm^{-3}}$

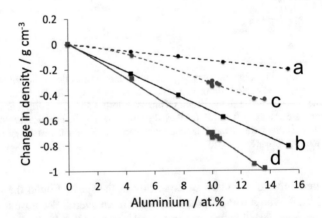

Fig. 11.3 (a) and (c) represent the changes in densities of ferrite and austenite, respectively, due
solely to lattice parameter expansion on alloying with aluminium. (b) and (d) represent the density
changes in ferrite and austenite, respectively, when both lattice parameter expansion and average
atomic mass changes due to alloying are taken into account. The lattice parameter data for ferrite
are from [11]. Those for austenite are from Fe-Mn-Al-C alloys [12] but with the lattice parameters
corrected so that the influence of Mn and C is removed

[13]. A recent alloy with similar density consists of a mixture of austenite and an intermetallic compound, with some of the manganese replaced in part with nickel (Fe-15Mn-5Ni-0.8C wt%) [16].

Like aluminium, silicon also stabilises ferrite at the expense of austenite. Fe-Si alloys containing about 4 wt% Si are ferritic at all temperatures below melting. Alloys containing up to 35 wt% Si have been studied [17, 18] but contain brittle intermetallic compounds at concentrations beyond about 17 wt% Si. Even when the silicon is completely in solid solution, the alloys tend to be brittle so concentrations greater than about 4 wt% are rarely used in order to control the density of steel.

Elements such as the alkali metals (Li, Na, K) and Be, Mg, Ca and Sc do not have sufficient solubility in solid steel to be useful, whereas Ti and V are strong carbide forming elements that can influence many other properties, not always in a beneficial manner.

Nitrogen is interesting — some 13 wt% can be introduced into the surface of austenitic steels either by diffusion or ion implantation [19, 20], causing a dramatic expansion in the lattice parameter of austenite by 10% resulting in a density reduction by an impressive 25%. Although the process results in outstanding surface properties, there is no practical method of introducing and retaining such large concentrations in bulk steel.

Figure 11.4a shows how the ultimate tensile strength normalised by density, compares for a variety of engineering metals. The data are plotted against elongation because many low-density steels are researched from the point of view of the

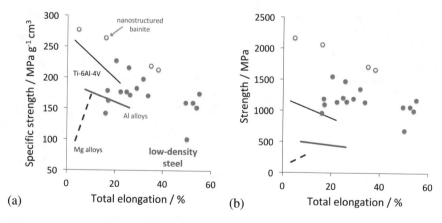

Fig. 11.4 (**a**) Specific ultimate tensile strength versus total elongation for steels [13, 14, 16, 22, 23], nanostructured bainitic steel [24], titanium and aluminium alloys [from compilation by 16], magnesium alloy [25]. (**b**) The same data but with the absolute ultimate strength on the ordinate

automotive industry where formability is an important factor in the selection of materials. It is evident that low-density steels outperform titanium, aluminium and magnesium alloys. The absolute strength levels of the low-density steels are plotted on Fig. 11.4b — it is important to realise that alloys whose strength exceeds about 1000 MPa cannot readily be pressed into components for automotive applications, because of a variety of factors such as elastic springback when the die is released. Nanostructured bainite and the stronger low-density steels would not therefore be suitable for formed components, although there will be other applications such as in shafts, etc. where formability is not an issue.

11.4   Concluding Remarks

In summary, it is possible to create alloys of iron that have a density much less than that of pure iron, with mechanical properties that are impressive. Now a bit of speculation — calculations based on electron theory can be used to estimate the energy of a crystal structure that does not exist on Earth. Figure 11.5 shows the cohesive energies of austenite (face-centred cubic) and the hypothetical diamond form of iron, at 0 K and zero pressure, as a function of the volume per atom [21]. If iron could be produced with the diamond structure, then its density would be a mere 5 g cm^{-3}!

Fig. 11.5 The cohesive energy versus the volume per atom (divided by that of an iron atom), for two allotropic forms of iron. Selected data from Paxton et al. [21]

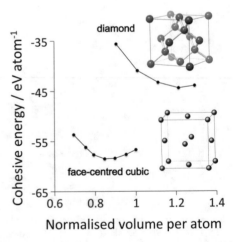

11.5 Terminology

Damper	A device or material that absorbs the energy due to externally applied stimuli. For example, can be used to dampen sound or vibrations.
Cohesive energy	The reduction in energy when atoms come together to form a crystal.
Electron theory	Here, a method of calculating energies as a function of atomic configurations.
Lattice parameter	The atoms in a crystal are arranged in a pattern that is a periodic array. The elementary, space-filling shape that represents the basic repeat-unit of the pattern is called a unit cell. The lengths of the edges of that cell are the lattice parameters.
Springback	When a metal is forced into a different shape by a press, the release of the pressure causes an amount of elastic relaxation that is known as the springback.

References

1. J.W. Arblaster, Crystallographic properties of osmium. Platin. Met. Rev. **57**, 177–185 (2013)
2. T.R. Hogness, The surface tensions and densities of liquid mercury, cadmium, zinc, lead, tin and bismuth. J. Am. Chem. Soc. **43**, 1621–1628 (1921)
3. L.E. Welch, D.M. Mossman, An environmental chemistry experiment. J. Chem. Educ. **71**, 521–523 (1994)
4. J.S. Lewis, Origin and composition of Mercury, in *Mercury*, ed. by F. Vilas, C. Chapman, M. Matthews (University of Arizona Press, Arizona, 1988), pp. 651–666.
5. F. Zwicky, On collapsed neutron stars. Astrophy. J. **88**, 522–525 (1938)
6. H.E. Cleaves, J.M. Hiegel, Properties of high-purity iron. J. Res. Natl. Bur. Stand. **28**, 643–667 (1942)
7. J.M. Singer, W.E. Wallace, The densities of magnesium-cadmium solid solutions. Den. Magnesium-Cadmium Solid Sol. **52**, 999–1006 (1948)
8. F.J. Brislee, The density of aluminium. Trans. Faraday Soc. **9**, 162–173 (1913)
9. K. Fujii, M. Tanaka, Y. Nezu, A. Sakuma, A. Leistner, W. Giardini, Absolute measurement of the density of silicon crystals in vacuum for a determination of the Avogadro constant. IEEE Trans. Instrum. Meas. **44**, 542–545 (1995)
10. E.J. Song, H.K.D.H. Bhadeshia, D.W. Suh, Interaction of aluminium with hydrogen in twinning-induced plasticity steel. Scr. Mater. **87**, 9–12 (2014)
11. A. Taylor, R.M. Jones, Constitution and magnetic properties of iron-rich iron-aluminum alloys. J. Phys. Chem. Solids **6**, 16–37 (1958)
12. C.M. Chu, H. Huang, P.W. Kao, D. Gan, Effect of alloying chemistry on the lattice constants of austenitic Fe-Mn-Al-C alloys. Scr. Met. Mater. **30**, 505–508 (1994)
13. G.L. Kayak, Fe-Mn-Al precipitation hardening austenitic alloys. Metalloved. Term. Obrab. Met. **112**, 13–16 (1969)
14. M.F. Alekseenko, G.S. Krivonogov, L.G. Kozyreva, I.M. Kachanova, L.V. Arapova, Phase composition, structure and properties of low-density steel 9G28Yu9MVB. Metalloved. Term. Obrab. Met. **14**(3), 2–4 (1972)
15. D.J. Schmatz, Formation of beta manganese-type structure in iron-aluminium-manganese alloys. Trans. Metall. Soc. AIME **215**, 112–114 (1959)

16. S.-H. Kim, H.S. Kim, N.J. Kim, Brittle intermetallic compound makes ultrastrong low-density steel with large ductility. Nature **518**, 77–79 (2015)
17. C.P. Yap, Critical study of some iron-rich iron-silicon alloys. J. Phys. Chem. **37**, 951–967 (1933)
18. K.M. Guggenheimer, H. Heitler, On phase-change processes in iron-silicon alloys. Trans. Faraday Soc. **45**, 137–145 (1949)
19. H. Dong, S-phase surface engineering of Fe-Cr, Co-Cr and Ni-Cr alloys. Int. Mater. Rev. **55**, 65–98 (2010)
20. D. Wu, H. Kahn, J.C. Dalton, G.M. Michal, F. Ernst, A.H. Heuer, Orientation dependence of nitrogen supersaturation in austenitic stainless steel during low-temperature gas-phase nitriding. Acta Mater. **79**, 339–350 (2014)
21. A.T. Paxton, M. Methfessel, H.M. Polatoglou, Structural energy-volume relations in first-row transition metals. Phys. Rev. B **41**, 8127–8138 (1990)
22. Y. Sutou, N. Kamiya, R. Umino, I. Ohnuma, K. Ishida, High-strength fe-20mn-al-c-based alloys with low density. ISIJ Int. **50**, 893–899 (2010)
23. G. Frommeyer, U. Brüx, Microstructures and mechanical properties of high-strength Fe-Mn-Al-C light-weight TRIPLEX steels. Steel Res. Int. **77**, 627–633 (2006)
24. H.K.D.H. Bhadeshia, *Bainite in Steels: Theory and Practice*, 3rd edn. (Maney Publishing, Leeds, 2015)
25. B.L. Mordike, T. Ebert, Magnesium – properties – applications – potential. Mater. Sci. Eng. A **302**, 37–45 (2001)

Chapter 12
Secrets of Ageless Iron Landmarks

12.1 Introduction

Metals and alloys have a tendency to degrade when exposed to the environment. In some cases, the corrosion products that form are attractive and even protect the underlying metal, in which case they are colloquially referred to as *patina*. Figure 12.1 shows two examples. The first is weathering steel, invented at U.S. Steel. It contains small concentrations of copper (0.25–0.55 wt%) and chromium (0.5–1.25 wt%), that help it develop a layer of patina. This layer of rust is more adherent and less porous than conventional rust which is not adherent. The major advantage is that little or no maintenance is required and the steel does not need to be painted at all. The lovely green copper patina is quite common in roofs of buildings or bronze statues. It is a copper oxide that develops over time.

However, in the vast majority of cases, the unintended reaction of metals with the fluids present in the environment is not desirable and costly to maintain or repair. Modern cars that are made of steel have highly sophisticated coating systems in place to mitigate corrosion. The turbine blades in the hot part of a jet engine would not survive the first operation without coatings that help them resist the aggressive environment within. In spite of all these efforts, corrosion costs countries around the world between 2–5% of the gross domestic product [1].

So it is surprising in this context that there are a number of iconic structures in existence, made of iron alloys, that have been exposed to the environment for many *hundreds* of years and yet remain remarkably intact.

The extent of degradation depends both on the alloy and its environment. Consider two well-visited iron landmarks. The Eiffel tower was erected in 1889 using iron before the advent of modern steelmaking processes, Fig. 12.2a. The intention was to dismantle it after 20 years, but today it remains the defining symbol of Paris, a must for any visitor to France. It has been painstakingly maintained to combat corrosion by an army of trained craftsmen. So its current spotless-condition and endurance are easy to understand. In contrast, an ancient structure known as the

© Springer Nature Switzerland AG 2021
T. DebRoy, H. K. D. H. Bhadeshia, *Innovations in Everyday Engineering Materials*,
https://doi.org/10.1007/978-3-030-57612-7_12

Fig. 12.1 (**a**) A footbridge made from weathering steel, located near the old Wall of London. Another example, not illustrated here is the fencing in the Rocky Mountains in Colorado which is all made of weathering steel; requires no maintenance. (**b**) A green patina of copper oxide on an ancient bronze cannon in Moscow. The statue of liberty is also covered in green copper patina

Fig. 12.2 (**a**) The Eiffel tower. Photograph courtesy of Hector Pous. (**b**) A photograph of the Delhi Pillar taken in 1872. The background and inscriptions are described in Cole [2]

iron pillar of Delhi, fabricated about 1600 years ago, has been exposed to oxygen and moisture in the air throughout its entire life. Surprisingly, this iron pillar is still largely intact, Fig. 12.2b. The tenacity of the Delhi pillar is intriguing and understanding the mechanism of its resistance to corrosion could be helpful in the modern context.

Many of the iconic structures are of cast or wrought iron. The cast irons manufactured more than a century ago contained typically 3.4 wt% of carbon, 2 wt% of silicon and other elements in small concentrations. Their structure contained

Fig. 12.3 (a) The Iron Bridge across the Severn River in England, completed in 1781 is the first large bridge to be made of cast iron, designed so that much of it is compression. (b) HMS Warrior, launched during 1861, was the first iron-hulled ship, built using wrought iron which is much more ductile than the cast iron of that age. The wrought iron also served as armour

(a)

(b)

flakes of graphite and hence the material was not strong in tension. Figure 12.3a shows the first large-scale bridge made of cast iron, with the arched shape allowing the members to be in compression, thereby stopping cracks from propagating.

Wrought iron, on the other hand, has a much smaller carbon concentration (≈ 0.1 wt%) so it does not contain graphite. It is malleable, can be forged and has good ductility. It can contain significant concentrations of sulphur (≈ 0.1 wt%) and phosphorus (≈ 0.2 wt%) and stringers of slag (oxides or silicates). Wrought iron can be bent and twisted so it has great utility in decorative window grills, fences and gates. The first warship made with a metallic hull used wrought iron which also served as armour (Fig. 12.3).

12.2 Manufacture

12.2.1 Wrought Iron

Pure iron melts at 1811 K which is too high a temperature for primitive furnaces constructed to handle the low melting temperatures of copper alloys towards the end of the bronze age. So it is logical to assume that ancient wrought iron was produced in the solid state at the beginning of the iron age, by reducing iron oxide to the solid iron.

When iron ore is heated in the presence of carbon and air, the oxide gets reduced to iron. This is because appreciable amounts of carbon monoxide is generated during the oxidation of carbon above 873 K, which in turn reacts with iron oxide to liberate the metallic iron. The iron so produced contains little carbon and is classified as wrought iron. It can easily be hammered into a variety of shapes at high temperatures. Wrought iron, used for thousands of years often contained pockets of impurities such as slags and small amounts of unreduced iron oxides trapped within the iron. As a result, the composition and structure of wrought iron is not as homogeneous as the modern steels and its properties vary depending on the method of manufacture and the impurity content.

12.2.2 Cast Iron

Cast iron, because of its large concentration of carbon, melts over a range of temperatures, typically about 1370–1470 K. The higher carbon-containing alloys melt at lower mean-temperatures and over a smaller temperature range. The lowest melting iron–carbon alloy, known as the *eutectic* alloy, contains 4.3% carbon and melts at 1420 K, in principle with a well-defined melting temperature rather than an extended range. Some primitive furnaces managed to attain temperatures somewhat higher than 1420 K and melted iron–carbon alloys that contained close to eutectic quantities of carbon. Therefore, cast iron has been produced in many countries for well over 2000 years, initially for making weapons and agricultural machinery and subsequently for industrial equipment such as the base for heavy machinery. The modern production of cast iron may use a combination of the iron from a blast furnace mixed with scrap iron and steel, coke, limestone and silicon, generally using electrically powered furnaces. And it is no longer the case that cast iron is brittle; it can be made so ductile that a bar can be repeatedly twisted without fracture.

12.3   Corrosion Mechanism

The physical processes that take place in most cases of corrosion, especially those connected with the iconic iron structures, are delightfully simple. There is in effect an electrical cell set-up between the dissolving surface (anode), the cathodic regions where oxygen in the electrolyte combines with water to produce hydroxyl ions that in turn combine with the metal ions evolving at the anodic site. The entire cell can exist in a droplet on the surface of iron, with the edges of the droplet being richer in oxygen becoming the cathodic sites and at the centre the anodic site. The first reaction in the corrosion of iron is:

$$Fe \rightarrow Fe^{2+} + 2e^{-1}$$

The ions thus generated need a medium to move away from the anode; this medium is an electrolyte which carries the ionic current. At the cathodic site, a common reaction that follows is:

$$\frac{1}{2}O_2 + H_2O + 2e^- \rightarrow 2OH^-$$

The reactions in the anode and cathode must be electrically equivalent. Finally, the electrons produced given up at the anode can be conducted through the iron towards the cathode where they are consumed. The iron ions react with the hydroxyl ions to produce $Fe(OH)_2$ which further oxidises to rust ($Fe(OH)_3$), but it is important to note that these reactions happen in the liquid, not at the surface of the iron. Therefore, the rust is not firmly attached to the iron and in this sense fails to protect it from further corrosion. The net reaction can be summarised as:

$$2Fe(OH)_2 + \frac{1}{2}O_2 + H_2O = 2Fe(OH)_3.$$

Details do depend on the alloy composition, for example, the rust may also contain other elements such as phosphorous. Unlike modern steels, ancient iron alloys typically contained significant amount of phosphorous. The importance of this will be described later.

When dissimilar metals are in contact in the presence of an electrolyte, an electrochemical or galvanic cell forms and the less noble metal forms the anode and corrodes locally at the contact. The *nobility* is a descriptive term for an electrode potential, which in turn is determined by the potential that develops between a metal and a reference electrode when both are immersed into an electrolyte of standard concentration. The metal with a lower electrode potential forms the anode, and corrodes. The prevention of galvanic corrosion involves separating the two metals by an electrical insulator so that the electrochemical cell does not form. When the separation is not practical, dissimilar metals have to be chosen carefully so that both metals have similar electrode potential so that the galvanic corrosion occurs very slowly.

Rusting of exposed iron surfaces often occurs slowly because it is uniform over large surfaces. However, in some cases corrosion occurs locally when there is a large imbalance between the anodic surface area and that which is cathodic. The corrosion current density at the anodic region is then exaggerated. For example, localised corrosion can begin in a tiny crevice where there may be a difference in the local oxygen concentration or in the nature of the electrolyte. Local differences in stresses in a metallic component may promote corrosion where the higher stressed region becomes the anode.

Prevention of corrosion may take several forms depending on its causative factors. Uniform corrosion over a large area can be mitigated by coating surfaces to form a barrier between the metal and the environment. Alternatively, a less noble metal coating can act as a sacrificial anode relative to which the iron is cathodic; this is how galvanising with zinc works.

12.4   Eiffel Tower: The Iron and Its Preservation

The Eiffel tower was erected to commemorate the centennial of the French Revolution. Soaring to a height of about 324 m, with its four pillars at the base spanning a 125 m square of 125 m each side, it was constructed from about 7300 tons of puddle iron, a form of wrought iron. Steel, which is a cleaner and more controlled form, was much more expensive than wrought iron at the time of its construction. The selection of puddle iron was the right economic choice for a structure that was designed to last only 20 years. Its unique shape and size, creative design, beauty and imposing presence has made it one of the most well-visited iron structures of all time, attracting some 7 million visitors during 2015. The optimism of the industrial revolution found a perfect expression in this iron structure.

The puddle iron was produced in small batches from pig iron made in the blast furnace (Chap. 7). Molten pig iron was oxidised in a hearth furnace to remove carbon and improve its properties. The heat necessary to melt the pig iron was generated in a fireplace with a cast iron grate where coal was burnt. The hot gases melted the pig iron without the coal coming in direct contact with the pig iron. Therefore, the iron was not contaminated by the impurities in the coal. Typically the charge started to melt in about 30 min after which the melt was stirred with long bars and a strong current of air was passed over it to oxidise carbon and other impurities in iron. At the end of the process, about 30–40 kg of a puddle ball of about 25–30 cm diameter was taken out using a large pair of tongs. The puddle iron was subsequently reheated and rolled to make bars or other products.

The puddle iron had fairly small concentrations of carbon and some other alloying elements such as manganese (Table 12.1). But most important, its structure contained streaks of slags not commonly found in steels and other alloys as shown in Fig. 12.4. The matrix is mostly soft ferrite. In the late nineteenth century, the

Table 12.1 Alloys used in Eiffel tower [3] and Delhi iron pillar[4, 5]

	Eiffel tower	Delhi pillar
Material	Puddle iron	Wrought iron
Method	Pig iron is melted in an open hearth furnace at the steelworks of Pompey. The product was hammered and rolled and made into thousands of metallic parts.	Ore heated in a charcoal fired furnace in solid state; sponge iron was hammer forged to about 20 to 30 kg lumps and multiple components were forge welded
Composition/wt%	0.05–0.25C, 0.05–0.2P, 0.02–0.1S, 0.02–0.2Si	0.15C, 0.25P, 0.005S, 0.05Si, 0.05Mn, 0.03Cu, 0.05Ni, 0.02N
Microstructure	Mostly ferrite, with small amounts of inclusions; sometimes small pearlite colonies near grain boundaries.	Mostly ferrite with pockets of slag inclusions and small amounts of unreduced iron oxides
Yield strength/MPa	207–448	323

Fig. 12.4 Microstructure of ancient wrought iron showing large slag inclusions. Reproduced from Pense [6], with permission of Elsevier

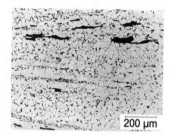

role of impurities such as phosphorous and sulphur were not well understood. As a result, the mechanical properties of the puddle iron were inferior than those of the steels then available. However, was this difference important for the structural integrity of the Eiffel tower? Since the structure was fairly light for its size and rested on a large foundation, the load on the foundation was safe and the main issue was the load bearing ability of the components. Much of the load from its own weight is compressive and rather light considering the lattice design of the tower. The exceptions are the torques produced by the wind and the non-uniform heating during summer. The curvatures of the four main pillars were calculated so that the tower does not sway much in the normal wind, typically about 5 cm or less. On a hot summer's day, the tower can move away from its normal position by about 14 cm due to the thermal expansion of the iron beams facing the sun.

The puddle iron parts were joined manually by riveting for the construction of the tower. After positioning the parts, the heated rivets were hammered into place and allowed to cool and contract, ensuring a firm joint. About 2,500,000 rivets were needed for the construction of the tower.

The greatest threat to the integrity of the iron beams is due to rusting in air. Gustave Eiffel recognised from the beginning the need for painting to resist corrosion. The iron parts are therefore stripped of the old paint, rust-proofed and a coat of paint applied manually every 7 years. The painting campaign also provides an opportunity to inspect the tower and replace any corroded parts. It takes a team of 25 painters about 18 months to paint the 200,000 m^2 of surface. The process uses 60 tonnes of paint, 1500 brushes and 1500 scraping abrasives. One of the campaigns started in April of 2009 and completed by the end of 2010 at a budget of about 4 million euros. Clearly paints have been effective in preventing any corrosion of the iron in the Eiffel tower.

12.5   Delhi Iron Pillar: Mysterious Corrosion Resistance

There is no record of any application of any protection measures with respect to corrosion, on this seven metre tall iron structure (Fig. 12.2b) and yet it has mysteriously defied pronounced rusting above ground for about 1600 years. Like many of the ancient iconic structures, it was made with wrought iron (Table 12.1). The pillar was made to honour the Hindu God Vishnu and celebrate a victory of King Chandra or Chandragupta II.

The pillar is coated by a relatively thick oxide film, 50–500 μm thick depending on the location. However, below ground, the pillar is covered by a considerable amount of rust, about 1 cm thick and there are several local deep oxidation pits. It is important to understand the nature of the rust because if the rust is impervious to oxygen and moisture and prevents access to the metal surface, further corrosion may be retarded. In addition, the initial exposure of the pillar to ammonia in the environment was also suggested as a contributing factor to the early passivation of the pillar against degradation.

It has been suggested that the rust contained crystalline iron hydrogen phosphate hydrate ($FePO_4 \cdot H_3PO_4 \cdot 4H_2O$) close to the metal layer and amorphous α-, γ- δ-FeOOH and magnetite outside the hydrogen phosphate hydrate layer [5] as shown in Fig. 12.5. The corrosion resistance of the pillar has been attributed to the hydrogen phosphate hydrate layer which largely protects iron from oxygen and moisture, thus significantly reducing the rate of corrosion. The formation of the crystalline iron hydrogen phosphate hydrate layer is attributed to alternate wetting and drying cycles because of the Delhi weather and the relatively high phosphorous content of the alloy. The schematic diagram in Fig. 12.5 on the left shows that the protective iron hydrogen phosphate hydrate layer does not form in the absence of phosphorous. Since the structure has survived over 1600 years and the retardation of corrosion has been systematically studied by multiple groups, what lessons can be learnt?

The roles of both the local environment and alloy composition were reviewed by Wranglen [4] who examined data on the atmospheric oxidation of small specimens of a carbon steel in various countries. The specimens were exposed to different environments for at least a year, cleaned and the loss of weight was measured in each case. The test results are shown in Table 12.2. The data show that the rates of atmospheric corrosion varied significantly as a function of location. The corrosion rate in Frodingham, England was about 100 times greater than that in Khartoum, Egypt. The corrosion rate in Delhi was significantly milder than in other locations

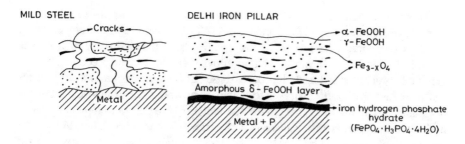

Fig. 12.5 Corrosion layers on mild steel and on the Delhi pillar. After Balasubramaniam [5], reproduced with the permission of Elsevier

Table 12.2 Corrosion rate of carbon steels in different locations [4]

Location	Climate	Thickness loss/μm year^{-1}	Relative corrosion rate
Khartoum, Egypt	Rural, desert-like	2	1
Delhi, India	Dry	5	3
Berlin, Germany	Urban	60	30
Pittsburgh, USA	Industrial	135	65
Frodingham, England	Industrial	200	100

tested, except for Khartoum. In addition, although some steels are clearly more susceptible to atmospheric oxidation than others, the steel to steel variations were not as pronounced as the 100-fold variations of oxidation rates in the locations shown in Table 12.2. The dry climate of Delhi, where the relative humidity does not exceed 70% for significant periods of the year, clearly was an important factor in the longevity of the pillar.

12.6 Concluding Remarks

The corrosion resistance of the Delhi iron pillar has been attributed to the formation of an adherent protective compound, iron-hydrogen-phosphate-hydrate, $FePO_4 \cdot H_3PO_4 \cdot 4H_2O$, which is thought to prevent the transport of oxygen and moisture to the inner core of the iron pillar. The dry climate of Delhi also helps reduce the corrosion rate. The protective layer forms because of the high phosphorous concentration, 0.25 wt% in the alloy used to build the pillar. Phosphorous is commonly present in iron ore and blast furnace pig iron but is largely eliminated into the slag during modern steelmaking. In most steels, the concentration of phosphorous now is limited to below 0.05 wt% because phosphorous decreases ductility and impact toughness and, under certain conditions, may cause the embrittlement of steels. Clearly, these properties are not of prime importance for the pillar.

Although corrosion resistance is important, it is not a single property that determines fitness-for-service of most alloys. This is because there are available a variety of methods for protecting the material from the environment. It now is possible to make a wide variety of iron alloys that can provide a combination of incredible properties at a fairly low cost. These outstanding properties of iron alloys and their low cost are the very reasons why they were used extensively from the beginning of the iron age (1500 BCE) and why they are poised to remain the most important engineering material for our civilisation in the foreseeable future. The cost of maintenance and prevention of corrosion remains high, mainly because we use such huge quantities of the iron alloys for the construction of buildings, bridges, ships, roads, cars and numerous other items that surround us.

12.7   Terminology

Ferritic steels	An iron–carbon alloy in which the unit cell consists of a cube with iron atoms at the corners and at the cube centre, commonly referred to as body-centred cubic with iron atoms occupying the corners of cubes and in the middle, commonly referred as body-centred cubic arrangement and carbon atoms positioned in the gaps of the iron atoms.
Forge welding	Process of solid-state welding where two alloys are heated and joined by application of force, typically by hammering.
Galvanic corrosion	Corrosion that happens when two dissimilar metals are in contact in the presence of an electrolyte.
Hearth furnace	A furnace where the charge rests on the hearth and heated by hot gases flowing over the charge.
Microstructure	A magnified view of different phases and defects in a specimen. Length scales that are observed are larger than those of atoms and molecules.
Passivation	Process of coating an alloy with a thin protective layer of adherent material to protect the alloy from corrosion.
Pig iron	Product of reducing iron ore in a blast furnace. Typically contains about 4.5 wt% carbon and other elements such as manganese, silicon, phosphorous and sulphur.
Solidus temperature	When a solid alloy is heated the highest temperature where the alloy is completely solid is called solidus temperature. Liquid formation starts when this temperature is exceeded.

References

1. M.V. Biezma, J.R.S. Cristóbal, Methodology to study the cost of corrosion. Corrosion Eng. Sci. Technol. **40**, 344–352 (2005)
2. H.H. Cole, *The Architecture of Ancient: Delhi, Especiallly the Buildings around the Kutb Minar* (The Arundale Society for Promoting the Knowledge of Art, London, 1919)
3. A.P.R. Company, *Wrought Iron and Steel in Construction* (Globe Printing House, Philadelphia, 1892)

4. G. Wranglén, The"rustless" iron pillar at Delhi. Corros. Sci. **10**, 761–770 (1970)
5. R. Balasubramaniam, On the corrosion resistance of the Delhi iron pillar. Corros. Sci. **42**, 2103–2129 (2000)
6. A.W. Pense, Iron through the ages. Mater. Charact. **45**, 353–363 (2000)

Index

© Springer Nature Switzerland AG 2021
T. DebRoy, H. K. D. H. Bhadeshia, *Innovations in Everyday Engineering Materials*,
https://doi.org/10.1007/978-3-030-57612-7